What We Believe *but Cannot Prove*

BOOKS BY JOHN BROCKMAN

AS AUTHOR

By the Late John Brockman

37

Afterwords

The Third Culture: Beyond the Scientific Revolution

Digerati

AS EDITOR

About Bateson

Speculations

Doing Science

Ways of Knowing

Creativity

The Greatest Inventions of the Past 2,000 Years

The Next Fifty Years: Science in the First Half of the 21st Century

The New Humanists: Science at the Edge

Curious Minds: How a Child Becomes a Scientist

AS COEDITOR

How Things Are

What We Believe *but Cannot Prove*

Today's Leading Thinkers on Science in the Age of Certainty

Edited by John Brockman

HARPER PERENNIAL

NEW YORK • LONDON • TORONTO • SYDNEY

TO MY PEERLESS LONGTIME EDITOR,
SARA LIPPINCOTT

HARPER PERENNIAL

HarperCollins books may be purchased for educational, business, or sales promotional use. For information please write: Special Markets Department, HarperCollins Publishers, 10 East 53rd Street, New York, NY 10022.

FIRST EDITION

Designed by Joy O'Meara

Library of Congress Cataloging-in-Publication Data is available upon request.

ISBN-10: 0-06-084181-8

ISBN-13: 978-0-06-084181-2

06 07 08 09 10 ❖/RRD 10 9 8 7 6 5 4 3 2

Contents

PREFACE

The *Edge* Question IX

INTRODUCTION

by Ian McEwan XIII

CONTRIBUTORS

Martin Rees 1

Ray Kurzweil 3

Douglas Rushkoff 7

Richard Dawkins 9

Chris Anderson 10

Stephen Petranek 11

Carolyn Porco 14

Paul C. W. Davies 17

Kenneth W. Ford 19

Karl Sabbagh 21

J. Craig Venter 23

Leon Lederman 25

Maria Spiropulu 27

Philip W. Anderson 29

Robert M. Sapolsky 30

Jesse Bering 32

Ian McEwan 36

Michael Shermer 37

Susan Blackmore 40
Randolph M. Nesse, M.D. 42
Tor Nørretranders 45
Scott Atran 47
David G. Myers 48
Jonathan Haidt 50
Sam Harris 51
David Buss 53
Seth Lloyd 55
Denis Dutton 56
Jared Diamond 59
Timothy Taylor 62
Judith Rich Harris 64
John H. McWhorter 68
Elizabeth Spelke 71
Stephen H. Schneider 73
Bruce Sterling 75
Robert Trivers 76
Verena Huber-Dyson 77
Keith Devlin 79
Freeman Dyson 82
Rebecca Goldstein 84
Stuart A. Kauffman 86
Leonard Susskind 88
Donald D. Hoffman 91
Terrence Sejnowski 95
John Horgan 98
Arnold Trehub 100
Ned Block 102
Janna Levin 103
Daniel Gilbert 105
Todd E. Feinberg, M.D. 107
Clifford Pickover 109

Nicholas Humphrey 111
Pamela McCorduck 113
Charles Simonyi 114
Alan Kay 118
Steven Pinker 120
Christine Finn 123
Daniel C. Dennett 124
Alun Anderson 128
Joseph LeDoux 132
George Dyson 136
Alison Gopnik 137
Paul Bloom 140
William H. Calvin 142
Robert R. Provine 145
Stanislas Dehaene 148
Stephen Kosslyn 152
Alex Pentland 154
Irene Pepperberg 158
Howard Gardner 161
David Gelernter 164
Marc D. Hauser 167
Gary Marcus 169
Brian Goodwin 171
Leo M. Chalupa 174
Margaret Wertheim 176
Gino Segrè 179
Haim Harari 181
Donald I. Williamson 184
Ian Wilmut 187
Daniel Goleman 189
Esther Dyson 192
James J. O'Donnell 195
Jean Paul Schmetz 197

Nassim Nicholas Taleb 199
Simon Baron-Cohen 201
Kevin Kelly 203
Martin Nowak 206
Tom Standage 207
Steven Giddings 209
Alexander Vilenkin 212
Lawrence M. Krauss 214
John D. Barrow 216
Paul J. Steinhardt 217
Lee Smolin 220
Anton Zeilinger 223
Gregory Benford 225
Rudy Rucker 227
Carlo Rovelli 229
Jeffrey Epstein 231
Howard Rheingold 232
Jaron Lanier 235
Marti Hearst 239
Kai Krause 241
Oliver Morton 244
W. Daniel Hillis 246
Martin E. P. Seligman 248
Neil Gershenfeld 250
Mihaly Csikszentmihalyi 251

Preface

The *Edge* Question

In 1991 I suggested the idea of a third culture, which "consists of those scientists and other thinkers in the empirical world who, through their work and expository writing, are taking the place of the traditional intellectual in rendering visible the deeper meanings of our lives, redefining who and what we are." By 1997 the growth of the Internet had allowed implementation of a home for the third culture on the Web, on a site named *Edge* (www.edge.org).

Edge is a celebration of the ideas of the third culture, an exhibition of this new community of intellectuals in action. They present their work, their ideas, and comment about the work and ideas of third culture thinkers. They do so with the understanding that they are to be challenged. What emerges is rigorous discussion concerning crucial issues of the digital age in a highly charged atmosphere where "thinking smart" prevails over the anesthesiology of wisdom.

The ideas presented on *Edge* are speculative; they represent the frontiers in the areas of evolutionary biology, genetics, computer science, neurophysiology, psychology, and physics. Some of the fundamental questions posed are: Where did the universe come from? Where did life come from? Where did the mind come from? Emerging out of the third culture is a new natural philosophy, new ways of understanding physical systems, new ways of thinking about thinking that call into question many of our basic assumptions of who we are, of what it means to be human.

An annual feature of *Edge* is The World Question Center, which was introduced in 1971 as a conceptual art project by my friend and collaborator the late artist James Lee Byars, who died in Egypt in 1997. I met Byars in 1969, when he sought me out after the publication of my first book, *By the Late John Brockman*. We were both in the art world, we shared an interest in language, in the uses of the interrogative, and in "the Steins"— Einstein, Gertrude Stein, Wittgenstein, and Frankenstein. Byars inspired the idea of *Edge* and is responsible for its motto:

> *To arrive at the edge of the world's knowledge, seek out the most complex and interesting minds, put them in a room together, and have them ask each other the questions they are asking themselves.*

He believed that to arrive at an axiology of societal knowledge it was pure folly to go to a Widener Library and read 6 million books. (He kept only four books at a time in a box in his minimally furnished room, replacing books as he read them.) His plan was to gather the 100 most brilliant minds in the world together in a room, lock them in, and "have them ask each other the questions they were asking themselves." The result was to be a synthesis of all thought. But between idea and execution are many pitfalls. Byars identified his 100 most brilliant minds, called each of them, and asked them what questions they were asking themselves. The result: seventy people hung up on him.

By 1997 the Internet and e-mail had allowed for a serious

implementation of Byars's grand design, and this resulted in launching *Edge*. Among the first contributors were Freeman Dyson and Murray Gell-Mann, two names on his 1971 list of the 100 most brilliant minds in the world.

For each of the eight anniversary editions of *Edge* I have used the interrogative myself and asked contributors for their responses to a question that comes to me, or to one of my correspondents, in the middle of the night. The 2005 *Edge* Question was suggested by the theoretical psychologist Nicholas Humphrey.

> *Great minds can sometimes guess the truth before they have either the evidence or arguments for it. (Diderot called it having the "esprit de divination.") What do you believe is true, even though you cannot prove it?*

The 2005 *Edge* Question was an eye-opener (BBC 4 Radio characterized it as "fantastically stimulating . . . the crack cocaine of the thinking world"). In the responses gathered here, there's a focus on consciousness, on knowing, on ideas of truth and proof. If pushed to generalize, I would say that these responses form a commentary on how we are dealing with a surfeit of certainty. We are in the age of search culture, in which Google and other search engines are leading us into a future rich with an abundance of correct answers along with an accompanying naive sense of conviction. In the future, we will be able to answer the questions—but will we be bright enough to ask them?

This book proposes an alternative path. It may be that it's OK not to be certain, but to have a hunch and to perceive on that basis. As Richard Dawkins, the British evolutionary biologist and champion of the public understanding of science, noted in an interview following publication of the 2005 Question: "It would be entirely wrong to suggest that science is something that knows everything already. Science proceeds by having hunches, by making guesses, by having hypotheses, sometimes inspired by poetic thoughts, by aesthetic thoughts even, and then science goes about trying to demonstrate it experimentally or observationally. And that's the beauty of science—that it has this imaginative stage but then it goes on to the proving stage, the demonstrating stage."

There is also evidence in this book that scientists and their intellectual allies are looking beyond their individual fields—still engaged in their own areas of interest but, more important, thinking deeply about new understandings of the limits of human knowledge. They are seeing our science and technology not just as a matter of knowing things but as a means of tuning into the deeper questions of who we are and how we know what we know.

I believe that the men and women of the third culture are the preeminent intellectuals of our time. But I can't prove it.

Introduction

Proof, whether in science, philosophy, criminal court or daily life, is an elastic concept, interestingly beset with all kinds of human weakness, as well as ingenuity. When jealous Othello demands proof that his young wife is deceiving him (when, of course, she is pure), it is not difficult for Iago to give his master exactly what he masochistically craves. For centuries, brilliant Christian scholars demonstrated by rational argument the existence of a sky-god, even while they knew they could permit themselves no other conclusion. The mother wrongly jailed for the murder of her children, on the expert evidence of a pediatrician, reasonably questions the faith of the courts in scientific proof concerning sudden infant death syndrome. When Penelope is uncertain whether the shaggy stranger who turns up in Ithaca really is her husband Ulysses, she devises a proof invoking the construction of their nuptial bed which would satisfy most of us but not many logicians. The precocious ten-year-old mathematician who exults in the proof that the angles of a triangle always add up to 180 degrees will discover before his first shave that in other mathematical schemes this is not always so. Very few of us know how to demonstrate that two plus two equals four in all circumstances. But we hold it to be true, unless we are unlucky enough to live under a political dispensation that requires us to believe the impossible; George Orwell in fiction as well as Stalin, Mao, Pol Pot, and various others in fact, have shown us how the answer can be five.

It has been surprisingly difficult to establish definitively what the truth is about any matter, however simple. It is always hard to get a grasp of one's own innate assumptions, and it was once perilous to challenge the wisdom of the elders or the traditions that had survived the centuries, and dangerous to incur the anger of the gods, or at least of their earthly representatives. Perhaps it was the greatest invention of all, greater than that of the wheel or agriculture, this slow elaboration of a thought system, science, that has disproof at its heart and self-correction as its essential procedure. Only recently, over this past half millennium, has some significant part of humankind begun to dispense with the kinds of insights supposedly revealed by supernatural entities, and to support instead a vast and disparate mental enterprise that works by accretion, dispute, refinement and occasional radical challenges. There are no sacred texts—in fact, a form of blasphemy has turned out to be useful. Empirical observation and proof are, of course, vitally important, but some science is little more than accurate description and classification; some ideas take hold, not because they are proved but because they are consonant with what is known already across different fields of study, or because they turn out to predict or retrodict phenomena efficiently, or because persuasive persons with powers of patronage hold them—naturally, human frailty is well represented in science. But the ambition of juniors and adversarial methods, as well as mortality itself are mighty enforcers. As one commentator has noted, science proceeds by funerals.

And again, some science appears true because it is elegant—it is economically formulated while seeming to explain a great deal. Despite the fulmination against it from the pulpit,

Darwin's theory of natural selection gained rapid acceptance, at least by the standards of Victorian intellectual life. His proof was really an overwhelming set of examples laid out with exacting care. A relatively simple idea made sense across a huge variety of cases and circumstances, a fact not lost on an army of Anglican vicars in country livings who devoted their copious free time to natural history. Einstein's novel description, in his theory of general relativity, of gravitation as a consequence not of the attraction between bodies according to their mass but of the curvature of spacetime generated by matter and energy, was enshrined in textbooks within a few years of its formulation. Steven Weinberg describes how, from 1919 onward, various expeditions by astronomers set out to test the theory by measuring the deflection of starlight by the sun during an eclipse. Not until the availability of radio telescopes in the early fifties were the measurements accurate enough to provide verification. For forty years, despite a paucity of evidence, the theory was generally accepted because, in Weinberg's phrase, it was "compellingly beautiful."

Much has been written about the imagination in science, of wild hunches born out, of sudden intuitive connections, and of benign promptings from mundane events (let no one forget the structure of benzene and Kekule's dream of a snake eating its tail) to the occasional triumph of beauty over truth. In James Watson's account, when Rosalind Franklin stood before the final model of the DNA molecule, she "accepted the fact that the structure was too pretty not to be true." Nevertheless, the idea still holds firm among us laypeople that scientists do not believe what they cannot prove. At the very least, we demand of them higher standards of evidence than we expect from literary critics, journalists, or priests. It is for this reason that the annual *Edge* Question—What do you believe that you cannot prove?—has generated so much public interest, for there appears to be a paradox here: those who stake their intellectual credibility on rigorous proof are lining up to declare their various unfalsifiable beliefs. Should not skepticism be the kissing cousin of science? Those very men and women who castigated us for our insistence on some cloudy notion that was not subject to the holy trinity of blind, controlled, and randomized testing are at last bending the knee to declare their faith.

The paradox, however, is false. As the Nobel Prize laureate Leon Lederman writes in his reply, "To believe something while knowing it cannot be proved (yet) is the essence of physics." This collection, mostly written by working scientists, does not represent the antithesis of science. These are not simply the unbuttoned musings of professionals on their day off. The contributions, ranging across many disparate fields, express the spirit of a scientific consciousness at its best—informed guesswork that is open-minded, free-ranging, and intellectually playful. Many replies offer versions of the future in various fields of study. Those readers educated in the humanities, accustomed to the pessimism that is generally supposed to be the mark of a true intellectual, will be struck by the optimistic tone of these pages. Some, like the psychologist Martin Seligman, believe we are not rotten to the core. Others even seem to think that the human lot could improve. Generally evident in these pages is an unadorned pleasure in curiosity. Is there life, or intelligent life, beyond Earth? Does time really exist? Is language a precondition of consciousness? Are cockroaches conscious? Is there a theory beyond quantum mechanics? Or indeed, do we gain a selective advantage from believing things we cannot prove? The reader will find here a collective expression of wonder at the living and inanimate world that does not have an obvious equivalent in, say, cultural studies. In the arts, perhaps lyric poetry would be a kind of happy parallel.

Another interesting feature is the prevalence here of what

E. O. Wilson calls "consilience." The boundaries between different specialized subjects begin to break down when scientists find they need to draw on insights or procedures in fields of study adjacent or useful to their own. The old Enlightenment dream of a unified body of knowledge comes a little closer when biologists and economists draw on one another's concepts; neuro-scientists need mathematicians, molecular biologists stray into the poorly defended territories of chemists and physicists. Even cosmologists have drawn on evolutionary theory. And everyone, of course, needs sophisticated computing. To address each other across their disciplines, scientists have had to abandon their specialised vocabularies and adopt a lingua franca—common English. The accidental beneficiary, of course, has been the common reader, who needs no acquaintance with arcane jargon to follow the exchanges. One consequence—and perhaps symbol—of this emerging synthesis in the scientific community has been the Edge website and its peculiarly heady intellectual culture. These pages represent only a small part of an ongoing and thrilling colloquium that is open to all.

—Ian McEwan

Great minds can sometimes guess the truth before they have either the evidence or arguments for it. (Diderot called it having the "esprit de divination"). What do you believe is true even though you cannot prove it?

Martin Rees

SIR MARTIN REES is a professor of cosmology and astrophysics and the master of Trinity College at the University of Cambridge. He holds the honorary title of Astronomer Royal and is also a visiting professor at Imperial College London and Leicester University. He is the author of several books, including *Just Six Numbers*, *Our Cosmic Habitat*, and *Our Final Hour*.

I believe that intelligent life may presently be unique to our Earth but has the potential to spread throughout the galaxy and beyond it—indeed, the emergence of complexity could be near its beginning. If the searches conducted by SETI (the Search for Extra-Terrestrial Intelligence) continue to come up with nothing, that would not render life a cosmic sideshow; indeed, it would be a boost to our self-esteem. Terrestrial life and its fate would be seen as a matter of cosmic significance. Even if intelligence is now unique to Earth, there's enough time ahead for it to permeate at least this galaxy and evolve into a teeming complexity far beyond what we can conceive.

There's an unthinking tendency to imagine that humans will be around in 6 billion years to watch the sun flare up and die. But the forms of life and intelligence that have by then emerged will surely be as different from us as we are from a bacterium. That conclusion would follow even if future evolution proceeded at the rate at which new species have emerged over the past 3.5 or 4 billion years. But posthuman evolution (whether of organic species or artifacts) will proceed far faster than the changes that led to human emergence, because it will be intelligently directed rather than the gradual outcome of Darwinian natural selection. Changes will drastically accelerate in the present century—through intentional genetic modifications, targeted drugs, perhaps even silicon implants in the brain. Humanity may not persist as a single species for longer than a few more centuries, especially if communities have by then become established away from Earth.

But a few centuries is still just a millionth of the sun's future lifetime—and the universe probably has a much longer future. The remote future is squarely in the realm of science fiction. Advanced intelligences billions of years hence might even create new universes. Perhaps they'll be able to choose what physical laws prevail in their creations. Perhaps these beings could achieve the computational ability to simulate a universe as complex as the one we perceive ourselves to be in.

My belief may remain unprovable for billions of years. It could be falsified sooner—for instance, we or our immediate posthuman descendants may develop theories that reveal inherent limits to complexity. But it's a substitute for religious belief, and I hope it's true.

Ray Kurzweil

RAY KURZWEIL is an inventor, entrepreneur, and principal developer of (among a host of other inventions) the first print-to-speech reading machine for the blind, the first text-to-speech synthesizer, the first CCD flat-bed scanner, and the first commercially marketed large-vocabulary speech recognition system. Recipient of the National Medal of Technology among many other honors, he is the author of several books, including *The Singularity Is Near: When Humans Transcend Biology*.

We will find ways to circumvent the speed of light as a limit on the communication of information.

We are expanding our computers and communication systems both inwardly and outwardly. Our chips' features are ever smaller, while we deploy increasing amounts of matter and energy for computation and communication. (For example, we're making a larger number of chips each year.) In one or two decades, we will progress from two-dimensional chips to three-dimensional self-organizing circuits built out of molecules. Ultimately we will approach the limits of matter and energy to support computation and communication.

As we approach an asymptote in our ability to expand inwardly (that is, using finer features), computation will continue to expand outwardly, using materials readily available on Earth, such as carbon. But we will eventually reach the limits of our planet's resources and will expand outwardly to the rest of the solar system and beyond.

How quickly will we be able to do this? We could send tiny self-replicating robots at close to the speed of light, along with electromagnetic transmissions containing the needed software. These nanobots could then colonize faraway planets.

At this point, we run up against a seemingly intractable limit: the speed of light. Although a billion feet per second may seem fast, the universe extends over such vast distances that this appears to represent a fundamental limit on how quickly an advanced civilization (such as we hope to become) can spread its influence.

There are suggestions, however, that this limit is not as immutable as it may appear. Physicists Steve Lamoreaux and Justin Torgerson of the Los Alamos National Laboratory have analyzed data from an old natural nuclear reactor that 2 billion years ago produced a fission reaction lasting several hundred thousand years in what is now West Africa. Analyzing radioactive isotopes left over from the reactor and comparing them with isotopes from similar nuclear reactions today, they determined that the physics constant α (alpha, also called the fine structure constant), which determines the strength of the electromagnetic force, apparently has changed since 2 billion years ago. The speed of light is inversely proportional to α, and both have been considered unchangeable constants. Alpha appears to have decreased by 4.5 parts out of 108. If confirmed, this would imply that the speed of light has increased. There are other studies with similar suggestions, and there is a tabletop experiment now under way at Cambridge University to test our ability to engineer a small change in the speed of light.

Of course, these results will need to be carefully verified. If they are true, it may hold great importance for the future of our civilization. If the speed of light has increased, it has presumably done so not just because of the passage of time but because certain conditions have changed. This is the type of scientific insight that technologists can exploit. It is the nature of engineering to take a natural, often subtle scientific effect and control it, with a view toward greatly leveraging and magnifying it. If the speed of light has changed due to changing circumstances, that cracks open the door just enough for the capabilities of our future intelligence and technology to swing the door wide open. That's the nature of engineering: Consider, for example, how we have focused and amplified the subtle properties of Bernoulli's principle (that air rushing over a curved surface has a slightly lower pressure than it does over a flat surface) to create the whole world of aviation.

If it turns out that we are unable to change the speed of light, we may nonetheless circumvent it by using wormholes (which can be thought of as folds of the universe in dimensions beyond the three visible ones) as shortcuts to faraway places. In 1935 Einstein and the physicist Nathan Rosen devised a way of describing electrons and other particles as tiny spacetime tunnels. Twenty years later, the physicist John Wheeler described these tunnels as "wormholes," introducing that term for the first time. His analysis of wormholes showed them to be fully consistent with the theory of general relativity, which describes space as essentially curved in another dimension.

In 1988 Caltech physicist Kip Thorne and his PhD students Michael Morris and Uri Yertsever described in some detail how such wormholes could be engineered. Based on quantum fluctuations, so-called empty space is continually generating tiny wormholes the size of subatomic particles. By adding energy and following other requirements of both quantum physics and general relativity (two fields that have been notoriously difficult to integrate), these wormholes could in theory be expanded to allow objects larger than subatomic particles to travel through them. Sending humans would not be impossible, but it would be extremely difficult; however, as I have pointed out, we only need to send nanobots plus information, which could go through wormholes measured in microns rather than meters. The computational neuroscientist Anders Sandberg estimates that a one-nanometer wormhole could transmit a formidable 10^{69} bits per second. Thorne, Morris, and Yertsever also describe a method, consistent with general relativity and quantum mechanics, that could establish wormholes between Earth and faraway locations quickly even if the destination were many light-years away.

Physicist David Hochberg and Vanderbilt University's Thomas Kephart point out that shortly after the Big Bang, gravity was strong enough to have provided the energy required to spontaneously create enormous numbers of self-stabilizing wormholes. A significant portion of them are likely to still be around and may be pervasive, providing a vast network of corridors that reach far and wide throughout the universe. It might be easier to discover and use these natural wormholes than to create new ones.

The point is that if there are even subtle ways around the speed-of-light limit, the technological powers that our future human-machine civilization will achieve will discover them and leverage them to great effect.

Douglas Rushkoff

DOUGLAS RUSHKOFF is a media analyst, author, and documentarian. His books include *Nothing Sacred, Media Virus, Get Back in the Box,* and the novels *Ecstasy Club* and *Exit Strategy.*

Though I can't prove it more than anecdotally or experientially, I believe that evolution has purpose and direction. To me it seems obvious, if absolutely unconfirmable, that matter is groping toward complexity. True enough, stresses and threats, ranging from time and friction to decomposition and predators, require objects and life-forms to achieve some measure of durability in order to sustain themselves. But this ability to survive seems to me more a means to an end than an end in itself.

Theology goes a long way toward imbuing substance and processes with meaning—describing life as "matter reaching toward divinity," or as the process by which divinity calls matter back to itself. But theologians mistakenly ascribe this sense of purpose to history rather than to the future. This is only natural, since the narrative structures we use to understand our world tend to have beginnings, middles, and ends. In order to experience the payoff at the end of the story, we need to see it as somehow built into the original intention of events.

It's also hard for people to contend with the likely possibility that we are simply overadvanced fungi and bacteria hurtling through a galaxy in cold, meaningless space. But just because our existence may have arisen unintentionally and without purpose doesn't preclude meaning or purpose from emerging as a result of our interaction and collaboration. Meaning may not be a precondition for humanity as much as a by-product of it.

It's important to recognize that evolution at its best is a team sport. As Darwin's later, lesser known but more important works contend, survival of the fittest is a law that applies not as much to individuals as to groups. Likewise, most great leaps forward in human civilization, from the formation of clans to the building of cities, have been feats of collaborative effort. Increased survival rates are as much a happy side effect of good collaboration as its purpose.

If we could stop thinking of "meaning" and "purpose" as artifacts of some divine creative act and see them instead as the yield of our own creative future, they become goals, intentions, and processes very much in reach rather than the shadows of childlike, superstitious mythology.

The proof is impossible, since this is an unfolding story. Like reaching the horizon, arrival merely necessitates more travel.

Richard Dawkins

THE EVOLUTIONARY BIOLOGIST Richard Dawkins is the Charles Simonyi Professor of Public Understanding of Science at Oxford University and a Fellow of the Royal Society. His books include *The Selfish Gene*, *Climbing Mount Improbable*, *A Devil's Chaplain*, and *The Ancestor's Tale*.

It is an established fact that all of life on this planet is shaped by Darwinian natural selection, which also endows it with an overwhelming illusion of "design." I believe, but cannot prove, that the same is true all over the universe, wherever life may exist. I believe that all intelligence, all creativity, and all design, anywhere in the universe, is the direct or indirect product of a cumulative process equivalent to what we here call Darwinian natural selection. It follows that design comes late in the universe, after a period of Darwinian evolution. Design cannot precede evolution and therefore cannot underlie the universe.

Chris Anderson

CHRIS ANDERSON is the editor-in-chief of *Wired* magazine.

The Intelligent Design movement has opened my eyes. I realize that although I believe that evolution explains why the living world is the way it is, I can't actually prove it. At least not to the satisfaction of the ID folk, who seem to require that every example of extraordinary complexity and clever plumbing in nature be fully traced back (not just traceable back) along an evolutionary tree to prove that it wasn't directed by an Invisible Hand. If the scientific community won't do that, then, the argument goes, they must accept a large red "Theory" stamp on the evolution chapter in the biology textbooks and the addition of chapters on alternative theories, such as "guided" evolution and creationism.

So, by this standard, virtually everything I believe in must now fall under the shadow of unproveability. This includes the belief that democracy, capitalism, and other market-driven systems (including evolution!) are better than their alternatives. Indeed, I suppose I should now refer to them as "the theory of democracy" and "the theory of capitalism" and accept the teaching of fascism and living Marxism as alternatives in high schools.

Stephen Petranek

STEPHEN PETRANEK is the editor-in-chief of *Discover* magazine.

I believe that life is common throughout the universe and that we will find another Earthlike planet within a decade.

The mathematics alone ought to be proof enough for most people: billions of galaxies with billions of stars in each galaxy and planets around many of those stars. These numbers suggest that the absence of life elsewhere in the universe is an improbable scenario. But there is more to the idea than good odds. We've so far found more than 150 planets, just by looking at nearby stars in our little corner of the Milky Way—results suggesting that there are uncountable numbers of planets in the Milky Way alone. Some are likely to be Earthlike, or at least Earth-sized, although the vast majority of those we've found are huge gas giants, which—like Jupiter and Saturn—are unlikely to harbor life.

Five recent developments suggest that the discovery of extraterrestrial life is not far off.

First, NASA's Mars Rover *Opportunity* found incontrovertible evidence that a briny sea once covered its landing site, in the Martian plain known as Meridiani Planum. Now the only question about whether or not life once existed on Mars is whether that sea—which spread across Meridiani Planum twice in Martian history—existed long enough for life to form. The *Phoenix* mission, scheduled to land on the northern polar water-ice cap in May 2008 and study the cap's history and interaction with the Martian atmosphere, may answer that question.

Second, in February 2005, scientists studying images from the *Mars Express Orbiter* announced evidence near the planet's equator of an existing frozen lake the size of Earth's North Sea.

Third, a team of astrophysicists reported in July 2004 that radio emissions from Sagittarius B2, a nebula near the center of the Milky Way, indicate the presence of aldehyde molecules, the prebiotic stuff of life. Aldehydes help form amino acids, the fundamental components of proteins. Some of the same scientists had previously reported clouds of other organic molecules in space, including glycolaldehyde, a simple sugar. Outer space is doubtless full of complex molecules—not just atoms—necessary for life. Comets in other solar systems could easily deposit such molecules on planets, as those in our solar system may have done on Earth.

Fourth, astronomers are beginning to find much smaller planets around other stars. In the summer of 2004 a team led by Barbara McArthur of the University of Texas at Austin's McDonald Observatory found a planet eighteen times the mass of Earth (roughly the mass of Neptune) orbiting 55 Cancri, a star about as large as our sun, with three known planets. Around the same time, a team in Portugal announced their discovery of a fourteen-mass planet orbiting mu Arae, another sunlike star—the second planet to be found there. These smaller planets are likely to be rock, not gas. "We're on our way to finding an extrasolar Earth," McArthur told reporters.

Fifth, astronomers are getting good not only at finding new planets but also at improving the resolution of their telescopes enough to see them. Extrasolar planets had hitherto been found only by detecting the wobble that their gravitational pull exerts on the parent star. Better optical telescopes—like the large binocular telescope on Mount Graham, near Tucson—are nearing completion. A European consortium is planning a 100-meter telescope for Chile. Improved resolution will allow astronomers to analyze a planet's spectrum to determine its composition and what's on its surface—like water. Water, we have also discovered recently, is abundant in space in large clouds between and near stars.

So everything life needs is out there. For it not to come together somewhere else, as it did on Earth, is wildly improbable. There are so-called Goldilocks zones ("Not too hot, not too cold, just right!") in galaxies—regions where life as we know it is most likely to evolve and survive. (There's too much radiation toward the center of the Milky Way, for example.) And there are almost countless galaxies to explore. This is the Golden Age of astrophysics, and we're going to find life out there.

Carolyn Porco

THE PLANETARY SCIENTIST Carolyn Porco is a veteran of NASA's planetary exploration program and the leader of the imaging team for the *Cassini* mission to Saturn. She is the creator/editor of the Web site www.ciclops.org, where *Cassini* images are posted, and is currently a senior research scientist at the Space Science Institute in Boulder, Colorado.

This is a treacherous question to ask and a trivial one to answer. Treacherous because the shoals between the written lines can be navigated by some to the conclusion that truth and religious belief develop by the same means and are therefore equivalent. To those unfamiliar with the process by which scientific hunches and hypotheses advance to the level of verifiable fact, and the exacting standards applied in that process, the work of the scientist might seem no different from that of the prophet or the priest. Nothing could be further from reality: The scientific method relies on the deliberate high-magnification scrutiny and criticism by other scientists of any mechanisms proposed to explain the natural world. Unlike religious dogma, no matter how fervently a scientist may believe that something is true, his or her belief is not accepted as a true description of reality until it passes every executable test. Nature is the final arbiter, and great minds are great only insofar as they can intuit the way nature works and are shown by subsequent examination and proof to be right.

That said, this is, for me, a trivial question to answer. Though no one has yet shown that life of any kind other than Earthly life exists in the cosmos, I firmly believe that it does. My justification for this belief is a common one; no strenuous exertion of the intellect or suspension of disbelief required.

Our reconstruction of the history of the early solar system and of the events that led to the origin of Earth and its moon and the development of life on our planet informs us that self-replicating organisms originated from inanimate materials in a very narrow window of time. The tail end of the accretion of the planets—a period known as the heavy bombardment—ended about 3.8 billion years ago, approximately 800 million years after Earth formed. That was the time of formation and solidification of the big impact basins we see on the moon—and the time of the last catastrophe-causing impacts on the surface of Earth. Not until then did the terrestrial surface environment settle down and become conducive to the development of living organisms.

The first appearance of life-forms on Earth—the oldest fossils we have discovered so far—occurred shortly after that, around 3.5 billion years ago or even earlier. The interval in between—only 300 million years and less than the time represented by the rock layers in the walls of the Grand Canyon—is a proverbial blink of the cosmic eye. Despite the enormous complexity of even the simplest biological forms and processes and the long and complicated chain of chemical events that must have occurred to evolve animated molecular structures from inanimate atoms, it seems an inevitable conclusion that Earthly life developed quickly, as soon as the coast was clear long enough to do so.

Evidence is gathering that the events that created the solar system, driven predominantly by gravity, are common and pervasive in our galaxy and, by inductive reasoning, in galaxies throughout the cosmos. The cosmos is very, very big. Consider the overwhelming numbers of galaxies in the visible cosmos alone, all the sunlike stars in those galaxies, the number of habitable planets likely to be orbiting those stars, and the ease with which life developed on our own habitable planet, and it seems increasingly likely that life itself is a fundamental feature of our universe, along with dark matter, supernovae, and black holes.

I believe we are not alone. But it doesn't matter what I think, because I can't prove it. It is so beguiling a question, though, that humankind is actively seeking the answer. The search for life and so-called habitable zones is becoming more and more the focus of our planetary explorations. We may soon discover life-forms under the ice on some moon orbiting Jupiter or Saturn, or decode the intelligible signals of an advanced, unreachably distant, alien civilization. That will be a singular day indeed! I only hope I'm still around when it happens.

Paul C. W. Davies

PAUL C. W. DAVIES is a professor of natural philosophy in the Australian Centre for Astrobiology at Macquarie University, Sydney. His research spans the fields of cosmology, gravitation, and quantum field theory, with particular emphasis on black holes, the origin of the universe, and the origin of life. He is the author of numerous books, the latest of which is *How to Build a Time Machine*. Awards include the 2002 Faraday Prize of the Royal Society and, for his contributions to the deeper implications of science, the 1995 Templeton Prize.

One of the biggest of the Big Questions of existence is, Are we alone in the universe? Science has provided no convincing evidence one way or the other. It is certainly possible that life began with a bizarre quirk of chemistry, an accident so improbable that it happened only once in the entire observable universe, and we are it. On the other hand, maybe life gets going wherever there are Earthlike planets. We just don't know, because we have a sample of only one. However, no known scientific principle suggests an inbuilt drive from matter to life. No known law of physics or chemistry favors the emergence of the living state over other states. Physics and chemistry are, as far as we can tell, "life blind."

Yet I don't believe that life is a freak event. I think the universe is teeming with it. I can't prove it; indeed, it could be that humankind will never know the answer for sure. If we find life in our solar system, it most likely got there from Earth (or vice versa) in rocks kicked off planets by comet impacts. And to go beyond the solar system is the stuff of dreams. The best hope is that we develop instruments sensitive enough to detect life on extrasolar planets from Earth orbit. But, while not impossible, this is a formidable technical challenge.

So why do I think we are not alone, when we have no evidence for life beyond Earth? Not for the fallacious popular reason: "The universe is so big that there must be life out there somewhere." Simple statistics shows this argument to be bogus. If life is in fact a freak chemical event, it would be so unlikely to occur that it wouldn't happen twice among a trillion trillion trillion planets. Rather, I believe we are not alone because life seems to be a fundamental, and not merely an incidental, property of nature. It is built into the great cosmic scheme at the deepest level and therefore is likely to be pervasive.

I make this sweeping claim because life has produced mind, and through mind, beings who do not merely observe the universe but have come to understand it through science, mathematics, and reasoning. This is hardly an insignificant embellishment of the cosmic drama, but a stunning and unexpected bonus. Somehow life is able to link up with the basic workings of the cosmos, resonating with the hidden mathematical order that makes it tick. And that's a quirk too far for me.

Kenneth W. Ford

THE PHYSICIST Kenneth W. Ford is the retired director of the American Institute of Physics and the author of *The Quantum World: Quantum Physics for Everyone.*

I believe that microbial life exists elsewhere in our galaxy.

I am not even saying "elsewhere in the universe." If the proposition I believe to be true is to be proved true within a generation or two, I had better limit it to our own galaxy. I will bet on its truth there.

I believe in the existence of life elsewhere because chemistry seems to be so life-striving and because life, once created, propagates itself in every possible direction. Earth's history suggests that chemicals create life given almost any old mix of substances that includes a bit of water and almost any old source of energy; further, that life spreads into every nook and cranny over a wide range of temperatures, acidity, pressure, light level, and so on.

Believing in the existence of *intelligent* life elsewhere in the galaxy is another matter. Good luck to the SETI people and applause for their efforts, but consider that microbes have inhabited Earth for at least 75 percent of its history, whereas intelligent life has been around for just the blink of an eye—perhaps 0.02 percent of Earth's history (and for nearly all of that time without the ability to communicate into space). Perhaps intelligent life will have staying power; we don't know. But we do know that microbial life has staying power.

Now to a supposition: that Mars will be found to have harbored life and harbors life no more. If this proves to be the case, it will be extraordinarily sobering for humankind, even more so than the view of our fragile blue ball from the moon; even more so than our removal from the center of the universe by Copernicus, Galileo, and Newton; perhaps even more so than the discovery of life elsewhere in the galaxy.

Karl Sabbagh

KARL SABBAGH is a British television producer and writer who started his career in BBC television and now runs Skyscraper Productions, which produces a range of documentary, music, and drama programs for broadcasters in the U.K. and the U.S.A. He is the author of several books, including *The Riemann Hypothesis*.

I believe that if there is intelligent life elsewhere in the universe, of whatever form, it will be familiar with the concept of numerical counting.

Some philosophers believe that pure mathematics is human-specific and that an entirely different type of mathematics may well emerge from a different type of intelligence—a type of mathematics that has nothing in common with ours and may even contradict it. But it is difficult to imagine any sort of intelligent life-form that would not need to count with numbers. The stars in the sky are discrete points and cry out to be counted by intelligent beings throughout the universe (at least the ones who can see).

Intelligent objects with boundaries surely want to be measured ("I'm bigger than you," "I need a size-312 overcoat"). But perhaps there are life-forms that don't have boundaries and are, say, continuously varying density changes in some Jovian sea.

Intelligent life might be disembodied—or at least lack a discrete body—and merely shift between various points in a solid material matrix, so that it would be impossible to distinguish one intelligent being from another. But sooner or later—whether it be to measure the passing of time, the magnitude of distance, or the density of one Jovian being compared with another—numbers will have to be used. And if numbers are used, 2 + 2 must always equal 4; the number of stars in the Pleiades brighter than magnitude 5.7 will always be 11, which will always be a prime number; and two measurements of the speed of light in any units in identical conditions will always be identical. Of course, the fact that I find it difficult to think of beings that won't need our sort of mathematics doesn't mean they don't exist, but that's what I believe without proof.

J. Craig Venter

THE VISIONARY GENOMIC RESEARCHER J. Craig Venter is founder and president of the J. Craig Venter Institute and the J. Craig Venter Science Foundation. The Venter Institute conducts basic research that advances the science of genomics, specializes in genomic medicine and environmental and synthetic genomics, and explores the ethical and policy implications of genomic discoveries and advances.

I believe that life is ubiquitous in the universe and that life on our planet Earth most likely is the result of a panspermic event. The panspermia idea was first raised by Svante Arrhenius, who thought terrestrial life might have been "seeded" by microorganisms from outer space, and elaborated on by the late Francis Crick, who speculated that the primordial microorganisms came here in a rocket ship sent by an alien civilization (an act Crick called "directed panspermia").

DNA, RNA, and carbon-based life will be found wherever we find water and look with the right tools. Whether we can prove we have discovered life depends on our ability to improve the remote sensing of faraway systems; this in turn depends on whether we survive as a species for a sufficient period of time. As we have seen recently in the "shotgun" DNA sequencing of microorganisms collected en masse from the Sargasso Sea, when we investigate life here on Earth with the new tools of DNA sequencing, we find it in great abundance in the microbial world. In sequencing the genetic code of organisms that survive in the extremes of 0°C to temperatures well above the boiling point of water, or in strong acidic or alkaline environments so caustic that they would rapidly dissolve human skin, we begin to understand the breadth of life. Possible indicators of panspermia are organisms such as the bacterium *Deinococcus radiodurans*, which can survive millions of rads of ionizing radiation and complete desiccation for years, perhaps millennia. Within hours of being reintroduced into an aqueous environment, these microbes can repair any DNA damage they may have incurred.

Our humancentric view of life is clearly unwarranted. From the millions of genes we are continually discovering in all organisms, we learn that a finite number of genes appear over and over again and could easily have evolved from a few microbes arriving on a meteor or on intergalactic dust.

Panspermia is how life is spread throughout the universe, and we are contributing to it here on Earth by launching billions of microbes into space.

Leon Lederman

LEON LEDERMAN, director emeritus of the Fermi National Accelerator Laboratory, received the Nobel Prize in physics in 1988 (with Melvin Schwartz and Jack Steinberger) "for the neutrino beam method and the demonstration of the doublet structure of the leptons through the discovery of the muon neutrino." He is the author of several books, including (with Dick Teresi) *The God Particle* and (with Christopher Hill) *Symmetry and the Beautiful Universe*.

My friend the theoretical physicist believed so strongly in string theory ("It must be true!") that he was called to testify in a lawsuit that pitted string theory against quantum loop gravity. The opposing lawyer was skeptical. "What makes you such an authority?" he asked.

"Oh, I am without question the world's most outstanding theoretical physicist," was the startling reply. It was enough to convince the lawyer to change the subject. However, when the witness came off the stand he was surrounded by protesting colleagues. "How could you make such an outrageous claim?" they asked. The theoretical physicist defended himself: "Fellows, you just don't understand! I was under oath!"

To believe something while knowing that it cannot be

proved (yet) is the essence of physics. Guys like Einstein, Dirac, Poincaré, etc., extolled the beauty of concepts, in a bizarre sense placing truth at a lower level of importance. There are enough such examples that I have resonated with the arrogance of my theoretical masters, who were in effect saying that God (a.k.a. the Master, *Der Alte*) may have, in her fashioning of the universe, made some errors in favoring a convenient truth over a breathtakingly wondrous mathematics. This inelegant lack of confidence in the creator has heretofore always proved hasty. Thus, when the long respected and beautiful law of mirror symmetry was violated by weakly interacting but exotic particles, our pain at the loss of simplicity and harmony was greatly alleviated by the discovery of the failure of particle-antiparticle symmetry. The connection was exciting because the simultaneous reflection in a mirror and change of particles to antiparticles seemed to restore a new more beautiful and more powerful symmetry: CP (charge conjugation/parity) symmetry. CP symmetry gave us a connection between space (mirror reflection) and electric charge. How silly of us to have lost confidence in the essential beauty of nature!

The renewed confidence remained, even when it turned out that CP is also imperfectly respected. "Surely," we now believe, "there is some spectacular new unforeseen splendor in store for us!" She will not let us down. This we believe, even though we can't prove it.

Maria Spiropulu

MARIA SPIROPULU is an experimental physicist at CERN, the European Organization for Nuclear Research, in Geneva.

I believe nothing to be true if it cannot be proved.

I'll take the *Edge* Question and make a pseudoinvariant transformation that makes it more apt. When Niels Bohr was asked what the complementary variable of truth (i.e., *Wirklichkeit*, or "reality") was, he replied, with no hesitation, "*Klarheit*" (clarity). With apologies to Bohr—and since neither truth nor clarity are quantum mechanical variables—real truth and comprehensive clarity should be simultaneously achievable, given rigorous experimental evidence.

So I will use "clarity" (as in "clear reality") in place of "truth." I will also invent equivalents for "proof" and "belief." Proof will be interchangeable with "experimental scientific evidence." "Belief" is trickier, given that it has to do with complex carbon-based life. It can be interchangeable with "theoretical assessment" or "commonsense assessment," depending on the scale and the available technology. In this process (no doubt a path full of pitfalls), I have cannibalized the original *Edge* Question to produce the following:

What do you either commonsensically or theoretically assess to be clearly real, even though you have no experimental scientific evidence for it?

Now, this is a difficult question. There are many theoretical assessments made of the explanation of natural phenomena at the extreme energy scales, from the subnuclear to the supercosmic, that possess a degree of clarity. But all of them are buttressed by the vast collection of conciliatory data that, scale by scale, express nature's works. This is the case even for string theory. So the answer is still "Nothing."

With regard to Bohr's complementarity, I would suggest that belief and proof are in some way complementary: If you believe something, then you don't need proof of it, and if you have proof, you don't need to believe. (I would assign the hard-core string theorists who do not really care about experimental scientific evidence to the first category).

But the *Edge* Question seems to be inviting predictions of the big things to come in science. In my field, even frameworks that explain the world by invoking extra dimensions of space are old news. As a matter of fact, we are preparing to confirm or exclude them with data. My hunch (and my wish) is that in the laboratory we will be able to segment spacetime so finely that gravity will be studied and understood in a controlled environment—and that gravitational particle physics will become a recognized field.

Philip W. Anderson

PHILIP W. ANDERSON, the 1977 Nobel laureate in physics, is emeritus professor of physics at Princeton University and an external professor at the Santa Fe Institute. His principal interests are in condensed matter physics, biophysics, neural nets, and complexity theory.

Is string theory a futile exercise in physics, as I believe it to be? It is an interesting mathematical specialty and has produced and will produce mathematics useful in other contexts, but it seems no more vital as mathematics than other areas of very abstract or specialized math, and doesn't on that basis justify the incredible amount of effort expended on it.

My belief is based on the fact that string theory is the first science in hundreds of years to be pursued in pre-Baconian fashion—that is, without any adequate experimental guidance. It proposes that nature is the way we would like it to be, rather than the way we see it to be, and it is improbable that nature thinks the same way we do.

The sad thing, as several young would-be theorists have explained to me, is that string theory is so highly developed that it's a full-time job just keeping up with it. That means that other avenues are not being explored by the bright, imaginative young people and alternative career paths are blocked.

Robert M. Sapolsky

ROBERT M. SAPOLSKY is a professor of biological sciences at Stanford University and of neurology at Stanford's School of Medicine. He is the author of *A Primate's Memoir.*

Well, of course, it is tempting to go for something like ". . . that the wheel, agriculture, and the Macarena were all actually invented by yetis." Or for the sophomoric pseudoironic logic twist of ". . . that every truth can eventually be proved." Or to draw myself up to my full height and intone, "Sir, we scientists believe in nothing that cannot be proved by the whetstone of science. Verily, our faith is our lack of faith"—and then stomp off in a lab coat and a huff.

The first two aren't worth the words and the third just isn't so, no matter how many times you read *Arrowsmith.* Scientists are subjective human beings operating in an ostensibly objective business, so there are probably lots of things we take on faith.

So mine will be a fairly simple, straightforward proposal of an unjustifiable belief: namely, that there is no God(s) or such a thing as a soul (whatever the religiously inclined mean by that word).

I'm very impressed, moved, by one approach of the people on the other side of the fence. These are the believers who argue that it would be a disaster—would be the very work of Beelzebub—for

God's existence to be proved. What good would religiosity be if it came with a transparently clear contract instead of requiring a leap of faith into an unknowable void?

My own inclination is to not believe without requiring proof. Mind you, it would be perfectly fine with me if there were a proof that there is no God. Some might view this as a potential public health problem, given the number of people who would then run damagingly amok. But there's no shortage of folks running amok already, thanks to their belief in God, so it wouldn't be any more of a problem. All things considered, such a proof would be a relief. Many physicists, especially astrophysicists, seem weirdly willing to go on about their communing with God in contemplating the Big Bang, but in my world of biologists, the God concept gets mighty infuriating when you spend your time thinking about, say, untreatably aggressive childhood leukemia.

Finally, just to undo any semblance of logic, I might even continue to believe that there is no God even if it were proved that there is. A religious friend of mine once remarked that the concept of God is useful, because you can berate God during the bad times. But it is clear to me that I don't need to believe there is a God in order to berate him.

Jesse Bering

JESSE BERING is an assistant professor of experimental psychology at the University of Arkansas whose research centers on the links between empirical cognitive science and classic themes from existential philosophy.

In 1936, shortly after the outbreak of the Spanish Civil War, Miguel de Unamuno, author of the classic existential text *Tragic Sense of Life*, died alone in his office of heart failure at the age of seventy-two.

Unamuno was no religious sentimentalist. As a rector and professor of Greek at the University of Salamanca, he was an advocate of rationalist ideals and even died a folk hero for having openly denounced Francisco Franco's fascist regime. He was, however, ridden with a spiritual burden that troubled him nearly all his life. It was the problem of death. Specifically, the problem was his own death and what, subjectively, it would be "like" for him afterward: "The effort to comprehend it causes the most tormenting dizziness," he wrote.

I've taken to calling this dilemma Unamuno's paradox, because I believe that it is a universal problem. It is, quite simply, the materialist understanding that consciousness is snuffed out by death coming into conflict with the human inability to simulate the psychological state of death. Adopting a parsimonious stance allows one to easily deduce that we as corpses cannot experience mental states, but this theoretical proposition can be justified only by a working scientific knowledge (i.e., that the nonfunctioning brain is directly equivalent to the cessation of the mind). By stating that psychological states survive death, or even alluding to this possibility, one is committing oneself to a radical form of mind-body dualism.

Consider how bizarre it truly is: Death is seen as a transitional event that unbuckles the body from its ethereal soul, the soul being the conscious personality of the decedent and the once animating force of the now inert physical form. This dualistic view sees the self as being initially contained in bodily mass, as motivating overt action during this occupancy, and as exiting or taking leave of the body at some point after the body's expiration. So what, exactly, does the brain do, if mental activities can exist independently of it? After all, as John Dewey put it, "mind" is a verb, not a noun.

And yet this radicalism is especially common. In the United States alone, as much as 95 percent of the population reportedly believes in life after death. How can so many people be wrong? Quite easily, if you consider that we're all operating with the same blemished psychological hardware. One is tempted to argue, as Freud did, that it's just people's desire for an afterlife that's behind this widespread conviction. But it would be a mistake to leave it at that, although there is convincing evidence that emotive factors can be powerful contributors to people's belief in life after death. Whatever our motivations for rejecting or endorsing the idea of an immaterial soul that can defy physical death, we would be unable to form any opinion at all on the matter were it not for our species' ability to differentiate unobservable minds from observable bodies.

But here's the rub. The materialist version of death is the ultimate killjoy null hypothesis. The epistemological problem of knowing what it is "like" to be dead can never be resolved. Nevertheless, I think Unamuno would be proud of recent scientific attempts to address the mechanics of his paradox. In a recent study, for example, I reported that when adult participants were asked to reason about the psychological abilities of someone who had just died in an automobile accident, even participants who later classified themselves as "extinctivists" (people who agreed with the statement "What we think of as the soul, or conscious personality, of a person ceases permanently when the body dies") nevertheless stated that the dead person knew he was dead (a feat demanding, of course, ongoing cognitive abilities). One young extinctivist's answer was almost comical: "Yeah, he'd know—because I don't believe in the afterlife. It's nonexistent. He sees that now." Try hard as he might to be a good materialist, this subject couldn't help but be a dualist.

How do I explain such findings? Like trying to reconstruct one's mental states during dreamless sleep, consciously representing a final state of nonconsciousness poses formidable, if not impassable, cognitive constraints. By relying on simulation strategies to derive information about the minds of dead agents, you would in principle be compelled to "put yourself in their shoes," which is an impossible task. Several decades ago, the developmental psychologist Gerald Koocher found that a group of children who were tested on death comprehension reflected on what it might be like to be dead "with references to sleeping, feeling 'peaceful,' or simply 'being very dizzy.'" More recently, my colleague David Bjorklund and I observed that younger children are more likely to attribute mental states to a dead agent than are older children—which is precisely the opposite pattern one would expect to find if the origins of such beliefs could be traced exclusively to cultural learning.

It seems that the default cognitive stance is reasoning that human minds are immortal; the steady accretion of scientific facts may throw this stance off a bit but, as Unamuno found out, even science cannot answer the Big Question. Don't get me wrong. Like Unamuno, I don't believe in the afterlife. Recent findings have led me to believe that it's all a cognitive illusion churned up by a psychological system specially designed to think about unobservable minds. The soul is distinctly human all right. Without our evolved ability to reason about minds, the soul would never have been. But in this case, the proof isn't in the empirical pudding. It can't be. It's death we're talking about, after all.

Ian McEwan

THE BRITISH NOVELIST Ian McEwan is the author of, among other books, *Enduring Love*, *Amsterdam*, *Atonement*, and *Saturday*.

What I believe but cannot prove is that no part of my consciousness will survive my death. I exclude the fact that I will linger, fadingly, in the thoughts of others, or that aspects of my consciousness will survive in writing, or in the positioning of a planted tree or a dent in my old car. I suspect that many contributors to *Edge* will take this premise as a given: true but not significant. However, it divides the world crucially, and much damage has been done to thought as well as to persons by those who are certain that there is a life—a better, more important life—elsewhere. That this span is brief, that consciousness is an accidental gift of blind processes, makes our existence all the more precious and our responsibilities for it all the more profound.

Michael Shermer

MICHAEL SHERMER is the publisher of *Skeptic* magazine, a columnist for *Scientific American*, and the author of *The Science of Good and Evil*.

I believe, but cannot prove, that reality exists independent of its human and social constructions. Science as a method, and naturalism as a philosophy, together form the best tool we have for understanding that reality. Because science is cumulative, building on itself in progressive fashion, we can achieve an ever greater understanding of reality. Our knowledge of nature remains provisional because we can never know if we have final Truth. Because science is a human activity and nature is complex and dynamic, fuzzy logic and fractional probabilities best describe both nature and our approximate understanding of it.

There is no such thing as the paranormal and the supernatural; there is only the normal and the natural and mysteries we have yet to explain.

What separates science from all other human activities is its belief in the provisional nature of all conclusions. In science, knowledge is fluid and certainty fleeting. That is the heart of its limitation. It is also its greatest strength. There are, from this ultimate unprovable assertion, three additional insoluble derivatives.

1. *There is no God, intelligent designer, or anything resembling the divinity as proffered by the world's religions* (although an extraterrestrial being of significantly greater intelligence and power than us would probably be indistinguishable from God).

After thousands of years of attempts by the world's greatest minds to prove or disprove the divine existence or nonexistence, with little agreement among scholars as to the divinity's ultimate state of being, a reasonable conclusion is that the God question can never be solved and that one's belief, disbelief, or skepticism finally rests on a nonrational basis.

2. *The universe is ultimately determined, but we have free will.*

As with the God question, scholars of considerable intellectual power for many millennia have failed to resolve the paradox of feeling free in a determined universe. One provisional solution is to think of the universe as so complex that the number of causes and the complexity of their interactions make the predetermination of human action pragmatically impossible. We can even assign a value to the causal net of the universe to see just how absurd it is to think we can get our minds around it fully. It has been calculated that in order for a computer in the far future of the universe to resurrect in a virtual reality every person who ever lived or could have lived (that is, every possible genetic combination to create a human), with all the causal interactions between them and their environment, it would need 10^{10} to the power of 123 (a 1 followed by 10^{123} zeros) bits of memory. Suffice it to say that no computer in the conceivable future will achieve this level of power; likewise, no human brain even comes close.

The enormity of this complexity leads us to feel as though we were acting freely as uncaused causers, even though we are actually causally determined. Since no set of causes we select as the determiners of human action can be complete, the feeling of freedom arises out of this ignorance of causes. To that extent, we may act as though we were free. There is much to gain, little to lose, and personal responsibility follows.

3. *Morality is the natural outcome of evolutionary and historical forces, not divine command.*

The moral feelings of doing the right thing (such as virtuousness) or doing the wrong thing (such as guilt) were generated by nature as part of human evolution. Although cultures differ on what they define as right and wrong, the moral feelings of doing the right or wrong thing are universal to all humans. Human universals are pervasive and powerful and include at their core the fact that we are by nature moral and immoral, good and evil, altruistic and selfish, cooperative and competitive, peaceful and bellicose, virtuous and nonvirtuous. Individuals and groups vary in the expression of such universal traits, but everyone has them. Most people, most of the time, in most circumstances, are good and do the right thing, for themselves and for others. But some people, some of the time, in some circumstances, are bad and do the wrong thing for themselves and for others.

As a consequence, moral principles are provisionally true, where they apply to most people, in most cultures, in most circumstances, most of the time. At some point in the last 10,000 years (most likely around the time of the advent of writing and the shift from bands and tribes to chiefdoms and states some 5,000 years ago) religions began to codify moral precepts into moral codes and political states began to codify moral precepts into legal codes.

In conclusion, I believe but cannot prove that reality exists and science is the best method for understanding it; that there is no God; that the universe is determined but we are free; that morality evolved as an adaptive trait of humans and human communities; and that ultimately all of existence is explicable through science.

Of course, I could be wrong. . . .

Susan Blackmore

SUSAN BLACKMORE is a freelance writer, lecturer, and broadcaster, and a visiting lecturer at the University of the West of England, Bristol. Her research interests include memes and the theory of memetics, evolutionary theory, consciousness, and meditation. She is the author of numerous books, including *The Meme Machine*; *Consciousness: An Introduction*; and *Consciousness: A Very Short Introduction*.

It is possible to live happily and morally without believing in free will. As Samuel Johnson said, "All theory is against freedom of the will; all experience for it." With recent developments in neuroscience and theories of consciousness, theory is even more against it than it was in his time. So I long ago set about systematically changing the experience. I now have no feeling of acting with free will, although the feeling took many years to ebb away.

But what happens? People say I'm lying! They say it's impossible and so I must be deluding myself in order to preserve my theory. And what can I do or say to challenge them? I have no idea—other than to suggest that other people try the exercise, demanding as it is.

When the feeling is gone, decisions just happen with no

sense of anyone making them, but then a new question arises—will the decisions be morally acceptable? Here I have made a great leap of faith (or, more accurately, this body and its genes and memes and the whole universe it lives in have done so). It seems that when people discard the illusion of an inner self who acts, as many mystics and Buddhist practitioners have done, they generally do behave in ways that we think of as moral or good. So perhaps giving up free will is not as dangerous as it sounds—but this too I cannot prove.

As for giving up the sense of an inner conscious self altogether—this is very much harder. I just keep on seeming to exist. But though I cannot prove it, I think it is true that I don't.

Randolph M. Nesse, M.D.

RANDOLPH M. NESSE, M.D., is a professor of psychiatry and pyschology at the University of Michigan and director of the Evolution and Human Adaptation Program in the university's Institute for Social Research. His primary research goal is "to discover how natural selection shaped capacities for mood and the mechanisms that regulate them." He is coauthor (with George C. Williams) of *Why We Get Sick: The New Science of Darwinian Medicine.*

I can't prove it, but I'm pretty sure that people gain a selective advantage from believing in things they can't prove. Those who are occasionally consumed by false beliefs do better in life than those who insist on evidence before they believe and act. Those who are occasionally swept away by emotions do better than those who calculate every move. These advantages have, I believe, shaped mental capacities for intense emotion and passionate beliefs, because they give a selective advantage in certain situations.

I'm not advocating irrationality or extreme emotionality. Many, perhaps even most, of the problems plaguing individuals and groups arise from actions based on passion. The Greek initiators and Enlightenment implementers recognized, correctly, that the world would be better off if reason displaced superstition and raw emotion. I have no interest in returning to that road; fundamentalism, for example, remains a severe threat to civilization. I am arguing, however, that if we want to understand these tendencies, we need to stop dismissing them as defects and start considering how they came to exist.

I arrived at this belief from studying game theory and evolutionary biology while also seeing psychiatric patients. Many patients are consumed by fears, sadness, and other emotions they find painful and senseless, while others are crippled by grandiose fantasies or bizarre beliefs. Then there are those with obsessive-compulsive personality. These patients do not suffer from obsessive-compulsive disorder; they do not wash or count all day long. Their obsessive-compulsive personality is instead characterized by hyperrationality. They are mystified by other people's emotional outbursts. They do their duty and expect that others will, too. They are of course often disappointed in this, a disappointment giving rise to frequent resentment. They trade favors according to the rules, and they can fathom neither genuine generosity nor spiteful hatred.

People who lack passions suffer several disadvantages. When social life results in situations that can be mapped onto game theory, regular predictable behavior is seen to be a strategy inferior to allocating actions randomly among the options. The angry person who might seek revenge is a force to be wary of, while a sensible opponent can be easily dealt with. The passionate lover obliterates the superior but all too practical offer of marriage.

It's harder to explain the disadvantages suffered by people who lack a capacity for faith, but consider the outcomes for those who wait for proof before acting compared with those who act on confident conviction. The great things in life are done by people who go ahead when going ahead seems senseless to others. Usually they fail—but sometimes they succeed.

Like nearly every other trait, tendencies toward passionate emotion and irrational conviction are most advantageous in some middle range. The optimum for modern life seems to me to be located closer to the rational side of the median, but there are advantages and disadvantages at every point along the spectrum. Making human life better requires us to understand these capacities, and to do that, we must seek their origins and functions. I cannot prove this is true, but I believe it is. This belief spurs my search for evidence that will either strengthen my conviction or, if I can discipline my mind sufficiently, convince me it is false.

Tor Nørretranders

TOR NØRRETRANDERS, a science writer, lecturer, and consultant based in Copenhagen, is the author of *The User Illusion: Cutting Consciousness Down to Size*.

I believe in belief—or rather, I have faith in having faith. Yet I am an atheist (or a "bright," as some would have it), so how can that be?

It is important to have faith, but not necessarily in God. Faith is important far beyond the realm of religion: having faith in oneself, in other people, in the existence of truth and justice. There is a continuum of faith, from the basic everyday trust in others to the grand devotion to divine entities.

Recent advances in behavioral sciences, such as experimental economics and game theory, demonstrate that having faith is a common human attitude toward the world. Faith is vital in human interactions; it is no coincidence that the anchoring of behavior in risky trust is emphasized in systems of thought as diverse as Søren Kierkegaard's existentialist Christianity and modern theories of bargaining behavior in economic interactions. Both stress the importance of inner, subjective conviction as the basis for action, the feeling of an inner glow. You might say that modern behavioral science is rediscovering the importance of faith—something that has been known to religions for a long time. I would argue that this rediscovery shows us that the very act of having faith can be decoupled from a belief in divine entities.

So here is what I have faith in: We have a hand backing us—not a divine foresight or control but the very simple and concrete fact that we are all survivors. We are the result of a long line of survivors, who lived long enough to have offspring. Amoebae, reptiles, mammals. We can therefore be confident that we are expert at survival. We have an inner wisdom inherited from millions of generations of animals and human beings—a knowledge of how to go about life. That does not in any way imply foresight or planning ahead on our behalf. It implies only that we have reason to trust our ability to deal with whatever challenges we meet. We have inherited such an ability.

We have no guarantee of eternal life, not at all. The enigma of death is still there, ineradicable. But the basic fact that we are still here, despite snakes, stupidity, and nuclear weapons, gives us reason to have confidence in ourselves and each other, to trust others, to trust life itself. To have faith. Because we are here, we have reason for having faith in having faith.

Scott Atran

THE ANTHROPOLOGIST Scott Atran is a director of research at the Centre National de la Recherche Scientifique in Paris and an adjunct professor of psychology, anthropology, natural resources, and the environment at the University of Michigan's Institute for Social Research. He is the author of *Cognitive Foundations of Natural History* and *In Gods We Trust*.

There is no God that has existence apart from people's thoughts of God. There is certainly no Being that can simply suspend the (nomological) laws of the universe in order to satisfy our personal or collective yearnings and whims—like a stage director called on to improve a play. But there is a mental (cognitive and emotional) process, common to science and religion, of suspending belief in what you see and take for obvious fact. Humans have a mental compulsion—perhaps a by-product of the evolution of a hypersensitive reasoning device to serve our passions—to situate and understand the present state of mundane affairs within an indefinitely extendable and overarching system of relations between hitherto unconnected elements. In any event, what drives humanity forward in history is this quest for nonapparent truth.

David G. Myers

DAVID G. MYERS is a professor of psychology at Hope College, in Holland, Michigan. He is the author of several books, including *What God Has Joined Together? A Christian Case for Gay Marriage.*

As a Christian monotheist, I start with two unproved axioms:

1. There is a God.
2. It's not me (and it's also not you).

Together, these axioms imply my surest conviction: that some of my beliefs (and yours) contain error. We are, from dust to dust, finite and fallible. We have dignity but not deity. And that is why I further believe that we should:

a. hold all our unproved beliefs with a certain tentativeness,
b. assess others' ideas with open-minded skepticism, and
c. freely pursue truth aided by observation and experiment.

This mix of faith-based humility and skepticism helped fuel the beginnings of modern science, and it has informed my own research and science writing. The whole truth cannot be found merely by searching our own minds, for there is not enough there. So we also put our ideas to the test. If they survive, so much the better for them; if not, so much the worse.

Within psychology, this "ever reforming" process has many times changed my mind, leading me now to believe, for example, that newborns are not so dumb, that electroconvulsive therapy often alleviates intractable depression, that America's economic growth has not improved our morale, that the automatic unconscious mind dwarfs the conscious mind, that traumatic experiences rarely get repressed, that most folks don't suffer low self-esteem, and that sexual orientation is not a choice.

Jonathan Haidt

JONATHAN HAIDT is an associate professor in the Department of Psychology at the University of Virginia. His primary research is on morality and emotion and how they vary across cultures.

I believe, but cannot prove, that religious experience and practice is generated and structured largely by a few emotions that evolved for other reasons—particularly awe, moral elevation, disgust, and attachment-related emotions. That's not a supposition likely to raise any eyebrows in this forum.

But I further believe (and cannot prove) that hostility toward religion is an obstacle to progress in psychology. Most human beings live in a world full of magic, miracles, saints, and constant commerce with divinity. Psychology at present has little to say about those parts of life; we focus instead on a small set of topics that are fashionable or particularly tractable with our favorite methods. If psychologists took religious experience seriously and tried to understand it from the inside, as anthropologists do in studying other cultures, I believe it would enrich our science. I have found religious texts and testimonials about purity and pollution essential for understanding the emotion of disgust and for helping me to see the breadth of moral concerns beyond harm, rights, and justice.

Sam Harris

SAM HARRIS is the author of *The End of Faith: Religion, Terror, and the Future of Reason,* winner of the 2005 PEN Award for First Nonfiction. He is a graduate in philosophy from Stanford University and is currently completing his doctorate in neuroscience at UCLA, studying the neural basis of belief, disbelief, and uncertainty with functional magnetic resonance imaging (fMRI).

Twenty-two percent of Americans claim to be certain that Jesus will return to Earth to judge the living and the dead sometime in the next fifty years. Another 22 percent believe that he is likely to do so. The question that most interests me, both scientifically and socially, is the question of belief itself. What does it mean, at the level of the brain, to believe that a proposition is true? The difference between believing and disbelieving a statement—your spouse is cheating on you; you've just won $10 million—is one of the most potent regulators of human behavior and emotion. The instant we accept a given representation of the world as true, it becomes the basis for further thought and action; rejected as false, it remains a string of words.

What I believe, though cannot yet prove, is that belief is a content-independent process. Which is to say that beliefs about God—to the degree that they really are believed—are the same as beliefs about numbers, penguins, tofu, or anything else. This is not to say that all of our representations of the world are acquired through language, or that all linguistic representations are on the same logical footing. And we know that different regions of the brain are involved in judging the truth of statements drawn from different content domains. What I do believe, however, is that the neural processes governing the final acceptance of a statement as "true" rely on more fundamental, reward-related circuitry in our frontal lobes—probably the same regions that judge the pleasantness of tastes and odors. Truth may be beauty, and beauty truth, in more than a metaphorical sense. And false statements may quite literally disgust us.

Once the neurology of belief becomes clear and it stands revealed as an all-purpose emotion arising in a wide variety of contexts (often without warrant), religious faith will be exposed for what it is: a humble species of terrestrial credulity. We will then have additional, scientific reasons to declare that mere feelings of conviction are not enough when it comes time to talk about the way the world is. The only thing that guarantees that (sufficiently complex) beliefs actually represent the world are chains of evidence and argument linking them *to* the world. Only on matters of religious faith do sane men and women regularly dispute this fact. Apart from removing the principal reason we have found to kill one another, a revolution in our thinking about religious belief would clear the way for new approaches to ethics and spiritual experience. Ethics and spirituality lie at the very heart of what is good about being human, but our thinking on both fronts has been shackled to the preposterous for millennia. Understanding belief at the level of the brain may hold the key to new insights into the nature of our minds, to new rules of discourse, and to new frontiers of human cooperation.

David Buss

DAVID BUSS is a professor in the Psychology Department of the University of Texas at Austin. His research interests include the evolutionary psychology of human mating strategies, conflict between the sexes, jealousy, homicide, and stalking. He is the author of *The Evolution of Desire* and *The Murderer Next Door: Why the Mind Is Designed to Kill.*

I believe in true love.

I've spent two decades of my professional life studying human mating. In that time, I've documented phenomena ranging from what men and women desire in a mate to the most diabolical forms of sexual treachery. I've discovered the astonishingly creative ways in which men and women deceive and manipulate each other. I've studied mate poachers, obsessed stalkers, sexual predators, and spouse murderers. But throughout this exploration of the dark dimensions of human mating, I've remained unwavering in my belief in true love.

While love is common, true love is rare, and I believe that few people are fortunate enough to experience it. The roads of ordinary love are well traveled and their markers are well understood—the mesmerizing attraction, the ideational obsession, the sexual afterglow, the often profound self-sacrifice, the desire to combine DNA. But true love takes its own course, through uncharted territory. It knows no fences, has no barriers or boundaries. It's difficult to define, eludes modern measurement, seems scientifically woolly. But I know true love exists. I just can't prove it.

Seth Lloyd

SETH LLOYD is a quantum mechanical engineer and a professor in the Department of Mechanical Engineering at the Massachusetts Institute of Technology, where he specializes in the design of quantum computers and quantum communications systems. He is the author of *Programming the Universe*.

I believe in science. Unlike mathematical theorems, scientific results can't be proved. They can only be tested again and again until only a fool would refuse to believe them.

I cannot prove that electrons exist, but I believe fervently in their existence. And if you don't believe in them, I have a high-voltage cattle prod I'm willing to apply as an argument on their behalf. Electrons speak for themselves.

Denis Dutton

THE PHILOSOPHER Denis Dutton is founder and editor of the highly regarded Web publication Arts & Letters Daily (www.aldaily.com). He teaches the philosophy of art at the University of Canterbury, New Zealand, writes widely on aesthetics, and is editor of the journal *Philosophy and Literature*.

In a 1757 essay, the philosopher David Hume argued that because "the general principles of taste are uniform in human nature," the value of some works of art might be essentially eternal. He observed that "the same Homer who pleased at Athens and Rome two thousand years ago is still admired at Paris and London." The works that manage to endure over millennia, Hume thought, do so precisely because they appeal to deep, unchanging features of human nature.

Some unique works of art—for example, Beethoven's Pastoral Symphony—possess this rare ability to excite the human mind across cultural boundaries and through historic time. I cannot prove it, but I think a small body of such works—by Homer, Bach, Shakespeare, Murasaki Shikibu, Vermeer, Michelangelo, Wagner, Jane Austen, Sophocles, Hokusai—will be sought after and enjoyed for centuries or millennia into the future. Though fashions and philosophies are bound to change, these works will remain objects of permanent value to human beings.

The epochal survivors of art are more than just popular. The majority of works of popular art today are not inevitably shallow or worthless, but they tend to be easily replaceable. In the modern mass-art system, artistic forms endure, while individual works drop away. Spy thrillers, romance novels, pop songs, and soap operas are daily replaced by more thrillers, romance novels, pop songs, and soap operas. In fact, the ephemeral nature of mass art seems more pronounced than ever: Most popular works are incapable of surviving even a year, let alone a couple of generations. It's different with art's classic survivors. Even if they began (as the works of Sophocles and Shakespeare did) as works of popular art, they have set themselves apart in their durable appeal. Nothing kills them; audiences keep coming to experience these original works for themselves.

Against the idea of permanent aesthetic values is cultural relativism, which is taught as the default orthodoxy in many university departments. Aesthetic values have been widely construed by academics as mere contingent reflections of local social and economic conditions. Beauty, if not exactly in the eye of the beholder, has been misconstrued as being in the eyes of the society, a conditioning that determines values of cultural seeing. Such veins of explanation often include no small amount of cynicism: Why do people go to the opera? Oh, to show off their furs. Why are they thrilled by famous paintings? Because famous paintings are worth millions. Beneath such explanations is a denial of intrinsic aesthetic merit.

Such aesthetic relativism is decisively refuted, as Hume understood, by the cross-cultural appeal of a small class of art objects over centuries. Mozart packs Japanese concerts halls, as Hiroshige does Paris galleries, while new productions of Shakespeare in every major language of the world are endless. And finally it is beginning to look as though empirical psychology is equipped to address the universality of art. For example, evolutionary psychology is being used by literary scholars to explain the persistent themes and plot devices in fiction. The rendering of faces, bodies, and landscapes in art is amenable to psychological investigation. The structure of musical perception is now open to experimental analysis as never before. Poetic experience can be elucidated by the insights of contemporary linguistics. None of this research promises a recipe for creating great art, but it can throw light on what we already know about aesthetic pleasure.

What's going on most days in New York's Metropolitan Museum of Art, and most nights at Lincoln Center, are aesthetic experiences that will be continually revived and relived by our descendants into an indefinite future. This makes the creation of a great artist as permanent an achievement as the discovery of a great scientist. That much I think I know. The question we should now ask is, What makes this possible? What is it about the highest works of art that gives them eternal appeal?

Jared Diamond

JARED DIAMOND, an evolutionary biologist and a professor of geography at UCLA, is the author of the Pulitzer Prize–winning *Guns, Germs, and Steel: The Fates of Human Societies* and *Collapse: How Societies Choose to Fail or Succeed*. His field experience includes, among projects in North America, South America, Africa, Asia, and Australia, twenty-one expeditions to New Guinea and neighboring islands to study the ecology and evolution of birds.

When did humans complete their expansion around the world? I'm convinced, but can't yet prove, that humans first reached the continents of North America, South America, and Australia only very recently—during or near the end of the last Ice Age. Specifically, I'm convinced that they reached North America around 14,000 years ago, South America around 13,500 years ago, and Australia and New Guinea around 46,000 years ago, and that within a few centuries of those dates humans were responsible for the extinction of most of the big animals of those continents.

Background to my conjecture is that there are now thousands of sites with undisputed evidence of human presence dating back millions of years in Africa, Europe, and Asia, but in the Americas and Australia there are no sites with even disputed evidence of human presence more than 100,000 years ago. In the Americas the undisputed evidence suddenly appears in all the lower forty-eight U.S. states around 14,000 years ago, at numerous South American sites soon thereafter, and at hundreds of Australian sites between 46,000 and 14,000 years ago. Evidence of most of the former big mammals of those continents—e.g., elephants, lions, and giant ground sloths in the Americas and giant kangaroos and one-ton Komodo dragons in Australia—disappears within a few centuries of those dates. The transparent conclusion: People arrived then, quickly filled up those continents, and easily killed off their big animals—animals that had never seen humans and let humans walk up to them, as animals in the Galapagos and Antarctica still do today.

But Australian and American archaeologists resist this obvious conclusion, for several reasons. They try hard to find convincing earlier sites because such discoveries are dramatic. Every year, discoveries of purportedly older sites are announced, and as the supporting evidence dissolves or remains disputed, we are left in an increasing flux of new claims and vanishing old claims—like the Hydra, who sprouted two heads for every one cut off. There are only a few sites in the Americas with evidence of human butchering of the extinct big animals and none at all in Australia and New Guinea; and indeed, one would expect to find very few such sites, among all the sites of natural deaths for hundreds of thousands of years, if hunting had ended within a few decades because the prey became extinct.

Every year, beginning graduate students in archaeology and paleontology working in Africa or Europe or Asia uncover sites displaying an undisputed ancient human presence. Every year, such discoveries are announced for the Americas and Australia, too—but none has ever met the evidentiary standards of the sites in Africa, Europe, or Asia. The big animals of the latter three continents survive because they had millions of years to learn fear of human hunters, whose skills evolved very slowly. The big animals of the Americas and Australia died out because their very first human encounters were with skilled, fully modern hunters.

To my mind, the case is already proved. How many more decades of unconvincing claims will it take to convince the hold-outs among my colleagues? I don't know. It makes for more compelling newspaper headlines to report, "Wow!! We've overturned the established paradigm of American archaeology!!" instead of "Ho hum, yet another reportedly paradigm-overturning discovery fails to hold up."

Timothy Taylor

THE BRITISH ARCHAEOLOGIST Timothy Taylor, author of *The Prehistory of Sex* and *The Buried Soul,* teaches in the Department of Archaeological Sciences at the University of Bradford, U.K., and conducts research on later prehistoric societies in Eurasia.

"All your life you live so close to the truth, it becomes a permanent blur in the corner of your eye, and when something nudges it into outline it is like being ambushed by a grotesque," wrote Tom Stoppard in *Rosencrantz and Guildenstern Are Dead.*

I believe, even though I cannot prove it, that cannibalism and slavery were both prevalent in human prehistory. Neither belief commands specialist academic consensus and each phenomenon remains highly controversial, their empirical "signatures" in the archaeological record being ambiguous and fugitive.

"Truth" and "belief" are uncomfortable words in scholarship. It is possible to define as true only those things that can be proved by certain agreed-upon criteria. In general, science does not believe in truth—or, more precisely, science does not believe in belief. Understanding is understood as the best fit to the data under current limits (both instrumental and philosophical) of observation. If science fetishized truth, it would be religion, which it is not. However, under the conditions Thomas Kuhn designated as "normal science"—as opposed to the intellectual ferment of paradigm shifts—most scholars are clearly involved in supporting what is in effect a religion. Their best guesses become fossilized as a status quo, and the status quo becomes an item of faith. So when a scientist tells you that "the truth is . . . ," it's time to walk away. Better to find a priest.

The current generation of archaeologists have been inclined to say that the truths about cannibalism and about slavery are that each has sharp historical limits and each is a more or less aberrant cultural phenomenon. This has been in part a reaction against real and perceived biasses in Victorian and imperialist accounts of "primitive savagery" and partly a laudable attempt to tighten up criteria of proof in order to counter vague suppositions and romantic myth-making about the past. Thus in only a small number of cases has either behavior been accepted as true beyond reasonable doubt. But I see the problem in the starting point.

If we shift our background expectations and assume that coercing the living to do one's bidding is perhaps the first form of property ownership ("the slavery latent in the family," to use Marx and Engels' telling phrase) and eating the dead (as many wild vertebrates do) makes sense in nutritional and competitive terms, then the archaeologist's duty is to empirically establish those times and places where slavery and cannibalism ceased to exist. The only reason we have hitherto insisted on proof-positive rather than proof-negative in relation to these phenomena is that both seem grotesque to us today, and we have a high opinion of our natural civility. This is the most interesting point. The focus of my attention is how culturally elaborated mechanisms of restraint and interpersonal respect emerged and allowed such refined scruples.

Judith Rich Harris

JUDITH RICH HARRIS is the author of *The Nurture Assumption*. She is an independent scholar and theoretician whose interests include evolutionary psychology, social psychology, developmental psychology, and behavioral genetics.

I believe, though I cannot prove it, that three—not two—selection processes were involved in human evolution.

The first two are familiar: natural selection, which selects for fitness, and sexual selection, which selects for sexiness.

The third process selects for beauty, but not sexual beauty—not adult beauty. The ones doing the selecting weren't potential mates, they were parents. Parental selection, we can call it.

What gave me the idea was a passage from a book titled *Nisa: The Life and Words of a !Kung Woman*, by the anthropologist Marjorie Shostak. Nisa was about fifty years old when she recounted to Shostak, in remarkable detail, the story of her life as a member of a hunter-gatherer group.

One of the incidents described by Nisa occurred when she was a child. She had a brother named Kumsa, about four years younger than herself. When Kumsa was around three and still nursing, their mother realized she was pregnant again She explained to Nisa that she was planning to "kill"—that is, abandon at birth—the new baby, so that Kumsa could continue to nurse. But when the baby was born, Nisa's mother had a change of heart. "I don't want to kill her," she told Nisa. "This little girl is too beautiful. See how lovely and fair her skin is?"

Standards of beauty differ in some respects among human societies; the !Kung are lighter-skinned than most Africans and perhaps they pride themselves on this feature. But Nisa's story provides an insight into two practices that used to be widespread and that I believe played an important role in human evolution: the abandonment of newborns that arrived at inopportune times (this practice has been documented in many human societies by anthropologists) and the use of aesthetic criteria to tip the scales in marginal cases.

Coupled with sexual selection, parental selection could have produced certain kinds of evolutionary changes very quickly, even if the heartbreaking decision of whether to rear or abandon a newborn was made in only a small percentage of births. The characteristics that could be affected by parental selection would have to be apparent even in a newborn baby. Two such characteristics are skin color and hairiness.

Parental selection can help to explain how the Europeans, who are descended from Africans, developed white skin over such a short period of time. In Africa, a cultural preference for light skin (such as Nisa's mother expressed) would have been counteracted by other factors that made light skin impractical. But in less sunny Europe, light skin may actually have increased fitness, which means that all three selection processes might have worked together to produce the rapid change in skin color.

Parental selection coupled with sexual selection can also account for our hairlessness. In this case, I very much doubt that fitness played a role; other mammals of similar size—leopards, lions, zebras, gazelles, baboons, chimpanzees, and gorillas—get along fine with fur in Africa, where the change to hairlessness presumably took place. I believe (though I cannot prove it) that the transition to hairlessness took place quickly, over a short evolutionary time period, and involved only *Homo sapiens* or its immediate precursor.

It was a cultural thing. Our ancestors thought of themselves as "people" and thought of fur-bearing creatures as "animals," just as we do. A baby born too hairy would have been distinctly less appealing to its parents.

If I am right that the transition to hairlessness occurred very late in the sequence of evolutionary changes that led to us, then this can explain two of the mysteries of paleoanthropology: the survival of the Neanderthals in Ice Age Europe and their disappearance about 30,000 years ago.

I believe, though I cannot prove it, that Neanderthals were covered with a heavy coat of fur and that *Homo erectus*, their ancestor, was as hairy as the modern chimpanzee. A naked Neanderthal could never have made it through the Ice Age. Sure, he had fire, but a blazing hearth couldn't keep him from freezing when he was out on a hunt. Nor could a deerskin slung over his shoulders, and there is no evidence that Neanderthals could sew. They lived mostly on game, so they had to go out to hunt often, no matter how rotten the weather. And the game didn't hang around conveniently close to the entrance to their cozy cave.

The Neanderthals disappeared when *Homo sapiens*, who by then had learned the art of sewing, took over Europe and Asia. This new species, descended from a southern branch of *Homo erectus*, was unique among primates in being hairless. In their view, anything with fur on it could be classified as "animal"—or, to put it more bluntly, game. Neanderthal disappeared in Europe for the same reason the woolly mammoth disappeared there: The ancestors of the modern Europeans ate them. In Africa today, hungry humans eat the meat of chimpanzees and gorillas.

At present, I admit, there is insufficient evidence either to confirm or disconfirm these suppositions. However, evidence to support my belief in the furriness of Neanderthals may someday be found. Everything we currently know about this species comes from hard stuff, like rocks and bones. But softer things, such as fur, can be preserved in glaciers, and the glaciers are melting. Someday a hiker may come across the well-preserved corpse of a furry Neanderthal.

John H. McWhorter

THE LINGUIST John H. McWhorter is a senior fellow at the Manhattan Institute and the author of several books, including *Defining Creole*.

Not long ago, researching the languages of Indonesia for an upcoming book, I happened to find out about a few very obscure languages spoken on one island that are much simpler than one would expect. Most languages are much more complicated than they need to be; they take on needless baggage over the millennia simply because they can. So, for instance, most languages of Indonesia have a good number of prefixes and/or suffixes. Their grammars often force the speaker to attend to nuances of difference between active and passive much more than a European language does, and so on.

But here were a few languages—Keo, Ngada, Rongga—that had no prefixes or suffixes at all. Nor did they have any tones, like many other languages. For one thing, languages that have been around forever and that have no prefixes, suffixes, or tones are very rare. But where we do find them, they are whole groups, related variations on one another. Here, though, was a handful of small languages that contrasted bizarrely with hundreds of surrounding relatives.

One school of thought on how language changes says that

this kind of thing just happens by chance. But my work has shown me that contrasts like this one are due to sociohistory. Saying that "naked" languages like these are spoken alongside ones as bedecked as, for example, Italian is rather like saying that kiwis are flightless just *because*, rather than because their environment divested them of the need to fly.

For months I scratched my head over those languages. Why just those few, when none of their relatives had taken that odd path of development? Why there?

So isn't it interesting that the island those languages are spoken on is none other than Flores, which had its fifteen minutes of fame last year as the site where skeletons of "little people" were found. Anthropologists have hypothesized that this was a different species of *Homo*. While the skeletons date back 18,000 years ago or more, local legend recalls "little people" living alongside modern humans—little people who had some kind of language of their own and could "repeat back" in the modern humans' language.

The legends suggest that the little people had only primitive language abilities, but we can't be sure here. To the untutored layman who hasn't taken any twentieth-century anthropology or linguistics classes, an incomprehensible language may well sound like mere babbling.

What I "know" (very tentatively) but cannot prove (yet) is this: The reason that Keo and Ngada and Rongga are such strangely streamlined languages is that an ancestor of these languages, just as complex as its family members tend to be today, was used as a second language by the "little people" and simplified by them. Just as our classroom French and Spanish avoids or streamlines a lot of the hard stuff, people who learn a language as adults usually do not master it entirely.

Specifically, I would hypothesize that the little people were gradually incorporated into the modern-human society on Flores over time—perhaps subordinated in some way—such that the modern-human children heard the little people's rendition of the modern-human language as often as they heard that of its native speakers.

This kind of process explains, for example, why Afrikaans is a slightly simplified version of Dutch. Dutch colonists took on Bushmen as herders and nurses, and their children heard second-language Dutch as often as they heard the native Dutch of their parents. Pretty soon this new kind of Dutch was everyone's everyday language, and Afrikaans was born.

Much has been made of the parallels between the evolution of languages and the evolution of animals and plants. I believe that one important difference is that while animals and plants can evolve toward simplicity as well as complexity depending on conditions, languages do not evolve toward simplicity in any significant overall sense unless there is some sociohistorical factor at work.

Languages are always drifting into being, like Russian or Chinese or Navajo. They become like Keo and Ngada—or Afrikaans; or creole languages, like Papiamentu and Haitian; or even, I believe, English—only because of the intervention of such factors as forced labor and population relocation. Maybe we can now add interspecies contact to the list!

Elizabeth Spelke

ELIZABETH SPELKE is a professor of psychology at Harvard and a member of the university's Laboratory for Developmental Studies, which investigates how infants and children perceive and reason about the world around them.

I believe, first, that all people have the same fundamental concepts, values, concerns, and commitments, despite our diverse languages, religions, social practices, and expressed beliefs. If Israelis and Palestinians, defenders and opponents of abortion, or Cambridge intellectuals and Amazonian jungle dwellers were to get beyond their surface differences, each would discover a vast terrain of common ground. Our common conceptual and moral commitments spring from the core cognitive systems that allow newborn infants to grow into competent participants in any human society.

Second, one of our shared core systems centers on a notion that is false—the notion that members of different human groups differ profoundly in their concepts and values. This notion leads us to interpret the superficial differences between people as signs of deeper differences. It has quite a grip on us: Many people would lay down their lives for perfect strangers from their own community while looking with suspicion at members of other communities. And all of us are apt to feel a special pull toward those who speak our language and share our ethnic background or religion, relative to those who don't.

Third, the most striking feature of human cognition stems not from our core knowledge systems but from our ability to rise above them. Humans can discover that their core conceptions are false and replace them with truer ones. This change has happened dramatically in the domain of astronomy. Core abilities to perceive, act on, and reason about the surface layout predispose us to believe that the earth is a flat, extended surface on which gravity exerts a downward force. This belief has been decisively overturned by the progress of science. Today every child who plays computer games or watches *Star Wars* knows that the earth is one sphere among many, and that gravity pulls all these bodies toward one another.

Together, my three beliefs suggest a fourth. If the cognitive sciences are given sufficient time, the claim of a common human nature eventually will be supported by evidence as strong and convincing as the evidence that the earth is round. As humans are bathed in this evidence, we will overcome our misconceptions of human differences. Ethnic and religious rivalries and conflicts will come to seem as pointless as debates over the turtles that our pancake Earth sits upon. Our common need for a stable and sustainable environment for all people will be recognized. But this fourth belief is conditional. Our species is caught in a race between the progress of science and the escalation of intergroup conflicts. Will humans last long enough for our science to win this race?

Stephen H. Schneider

THE CLIMATOLOGIST Stephen H. Schneider is a professor in the Department of Biological Sciences at Stanford University and the codirector of its Center for Environmental Science and Policy. He is the author of several books, including *Global Warming* and *Laboratory Earth*.

I believe that global warming is both a real phenomenon and at least partly a result of human activities, such as pouring greenhouse gases into the atmosphere. In fact, I can prove it—or can I? That is the real question.

What is "proof"? In the strict, old-fashioned, frequentist statistical belief system, data consists of direct observations of the hypothesized phenomenon—temperature increase, in my case—and when you get enough to produce frequency distributions, you can assign objective probabilities to cause-and-effect hypotheses. But what if the events cannot be precisely measured—or worse, apply to future events, like the warming of the late twenty-first century? Then a frequentist interpretation of "proof" is impossible in principle, and we instead become subjectivists ("Bayesian updaters," as some statisticians like to refer to it), using frequency data and all other data relevant to components of our analysis to form a "prior"—that is, a belief about the likelihood of an event or process. Then, as we learn more, we update our belief ("an a-posteriori probability," the Bayesians call it).

It is my strong belief that there is an overwhelming amount of evidence—enough to form, with high confidence, a subjective prior that the earth's surface has warmed over the past century about 0.7° C or so and that at least half of the more recent warming is traceable to human pressures. Is this proof of anthropogenic (i.e., we did it) warming? Not in the strict sense of a criminal trial with the beyond-a-reasonable-doubt criterion—say, a 99-percent objective probability. But in the sense of a civil proceeding, where preponderance of evidence is the standard and a likelihood greater than 50 percent is enough to win the case, then global warming is indeed already proved. So, as a frequentist, I concede that I believe global warming is real without full proof, but as a subjectivist my reading of the many lines of evidence puts global warming well over the minimum threshold of belief—far enough to assert that it is already proved to the point where we need to consider taking it seriously.

Bruce Sterling

BRUCE STERLING is a novelist, journalist, and futurist. He is the author of *Tomorrow Now: Envisioning the Next 50 Years* and *The Zenith Angle*, a novel.

I can sum up my intuition in five words: We're in for climatic mayhem.

Robert Trivers

ROBERT TRIVERS, a professor of anthropology and biological sciences at Rutgers University, works in two areas: social theory based on natural selection (of which a theory of self-deception is one part) and, more recently, on "selfish" genetic elements that lead to internal genetic conflict within individuals. He is the author of *Natural Selection and Social Theory*.

I believe that deceit and self-deception play a disproportionate role in human-generated disasters: wars; misguided social, political, and economic policies; miscarriages of justice; the collapse of civilizations.

I believe that deceit and self-deception play an important role in the relative underdevelopment of the social sciences.

I believe that processes of self-deception are important in limiting the achievement of individuals.

Verena Huber-Dyson

VERENA HUBER-DYSON is a mathematician who has published research in group theory and has taught in several mathematics departments, including those at UC Berkeley and the University of Illinois at Chicago. She is now emeritus professor in the Philosophy Department of the University of Calgary, where she taught courses in logic and in the philosophy of the sciences and mathematics. She is the author of *Goedel's Theorems*.

Most of what I believe I cannot prove, simply for lack of time and energy—truths I'd claim to know because they have been proved by others. That is how inextricably our beliefs are tied up with labors accomplished by fellow beings. But since Goedel's proof of 1931, we know that the limitations of what can be proved are inherent in the concept of proof, not just in the limitations of the human mind. In fact, to every formal system, satisfying some natural minimal requirements, there exists a mathematical truth expressible in the system's language but not provable by its proof procedure. These phenomena have become favorites with the media but can be made sense of only by a serious scrutiny of the idea of mathematical truth and a specific articulation of a proof-concept.

But I can come up with a catchier response:

I believe in the creative power of boredom. Or, to put it into the form suggested by the *Edge* Question: I believe that no matter how relentlessly we overfeed our young with packaged, interactive entertainments, before long they will break out and invent their own amusements. I know from experience: Boredom drove me into mathematics during my preteens. But I cannot prove it until it actually happens. Probably in less than a generation, kids will be amusing themselves and each other in ways we never dreamed of. Such is my belief in human nature, in the resilience of its good sense.

Keith Devlin

THE MATHEMATICIAN Keith Devlin is executive director of Stanford University's Center for the Study of Langage and Information and a consulting professor in the Mathematics Department. His current research focuses on the design of information/reasoning systems for intelligence analysis. He is the author of several books, including *The Math Instinct: Why You're a Mathematical Genius (along with Lobsters, Birds, Cats, and Dogs)*.

Before we can answer the *Edge* Question, we need to agree what we mean by proof. (Mathematicians like to begin by giving precise definitions of what we are going to talk about, a pedantic tendency that sometimes drives our physicist and engineering colleagues crazy.) For instance, following Descartes, I can prove to myself that I exist, but I can't prove it to anyone else. Even to those who know me well, there is always the possibility, however remote, that I'm merely a figment of their imagination. If it's rock solid certainty you want from a proof, there's almost nothing beyond our own existence (whatever that means and whatever we exist as) that we can prove to ourselves, and nothing at all that we can prove to anyone else.

Mathematical proof is generally regarded as the most certain form of proof there is, and in the days when Euclid was writing his great geometry text *Elements* that was surely true in an ideal sense. But many of the proofs of geometric theorems that Euclid gave were subsequently found to be incorrect—David Hilbert corrected many of them in the late nineteenth century, after centuries of mathematicians had believed them and passed them on to their students—so even in the case of a ten-line proof in geometry, it can be hard to tell right from wrong.

When you look at some of the proofs that have been developed in the last fifty years or so, using incredibly complicated reasoning that can stretch into hundreds of pages or more, certainty is even harder to maintain. Most mathematicians (including me) believe that Andrew Wiles proved Fermat's last theorem in 1994, but did he really? I believe it because the experts in that branch of mathematics tell me they do.

In late 2002 the Russian mathematician Grigori Perelman posted on the Internet what he claimed was an outline for a proof of the Poincaré conjecture, a famous century-old topological problem. After examining the argument for three years now, mathematicians are still unsure whether it's right or not. (They think it "probably is.")

Or consider Thomas Hales, who is still waiting to hear whether or not the mathematical community accepts his 1998 proof of Johannes Kepler's 360-year-old conjecture that the most efficient way to pack equal-size spheres (such as cannonballs on a ship, which is how the question arose) is to stack them in the familiar pyramid fashion that greengrocers use to stack oranges on a counter. After examining Hales's argument (part of which was carried out by computer) for five years, a panel of world experts declared in spring of 2003 that whereas they had not found any irreparable error in the proof, they were still not sure it was correct.

With the idea of proof so shaky even in mathematics, answering this year's *Edge* Question becomes a tricky business.

The best we can do is come up with something we believe but cannot prove to our own satisfaction. Others will accept or reject what we say, depending on how much credence they give us as a scientist, a philosopher, or whatever, generally basing that decision on our reputation and the record of our previous work. Even the old mathematicians' standby of Gödel's incompleteness theorem (which on first blush would allow me to answer the *Edge* Question with a statement of my belief that arithmetic is free of internal contradictions) is no longer available. Gödel's theorem showed that you cannot prove that an axiomatically based theory like arithmetic is free of contradiction *within that theory itself*. But that doesn't mean you can't prove it in some larger, richer theory. In fact, in the standard axiomatic set theory, you *can* prove that arithmetic is free of contradictions. And personally, I buy that proof. For me, as a living, human mathematician, the consistency of arithmetic has been proved to my complete satisfaction.

So to answer the *Edge* Question, you have to take a commonsense approach to proof—in this case, proof being an argument that would convince the intelligent, professionally skeptical, trained expert in the appropriate field. In that spirit, I could give any number of specific mathematical problems that I believe are true but cannot prove, starting with the famous Riemann hypothesis. But I think I can be of more use by using my mathematician's perspective to point out the uncertainties in the idea of proof. Which I believe (but cannot prove) I have.

Freeman Dyson

FREEMAN DYSON is professor emeritus of physics at the Institute for Advanced Study, in Princeton. He is the author of a number of books about science for the general public, including *Imagined Worlds* and *The Sun, the Genome, and the Internet*.

Since I am a mathematician, I give a precise answer to this question. Thanks to Kurt Gödel, we know that there are true mathematical statements that cannot be proved. But I want a little more than this. I want a statement that is true, unprovable, and simple enough to be understood by people who are not mathematicians. Here it is.

Numbers that are exact powers of two are 2, 4, 8, 16, 32, 64, 128, and so on. Numbers that are exact powers of five are 5, 25, 125, 625, and so on. Given any number, such as 131,072 (which happens to be a power of two), the reverse of it is 270,131, with the same digits taken in the opposite order. Now, my statement is: It never happens that the reverse of a power of two is a power of five.

The digits in a big power of two seem to occur in a random way, without any regular pattern. If it ever happened that the reverse of a power of two was a power of five, this would be an unlikely accident, and the chance of it happening grows rapidly smaller as the numbers grow bigger. If we assume that the digits occur at random, then the chance of the accident happening for any power of two greater than a billion is less than one in a billion. It is easy to check that it does not happen for powers of two smaller than a billion. So the chance that it ever happens at all is less than one in a billion. That is why I believe the statement is true.

But the assumption that digits in a big power of two occur at random also implies that the statement is unprovable. Any proof of the statement would have to be based on some nonrandom property of the digits. The assumption of randomness means that the statement is true just because the odds are in its favor. It cannot be proved, because there is no deep mathematical reason why it has to be true. (Note for experts: This argument does not work if we use powers of three instead of powers of five. In that case the statement is easy to prove, because the reverse of a number divisible by three is also divisible by three. Divisibility by three happens to be a nonrandom property of the digits).

It is easy to find other examples of statements that are likely to be true but unprovable. The essential trick is to find an infinite sequence of events, each of which might happen by accident, but with a small total probability for even one of them happening. Then the statement that none of the events ever happens is probably true but cannot be proved.

Rebecca Goldstein

REBECCA GOLDSTEIN is a novelist and a professor of philosophy at Trinity College in Hartford, Connecticut. She is the author of *Incompleteness: The Proof and Paradox of Kurt Gödel* and six works of fiction, including *The Mind-Body Problem* and *Properties of Light: A Novel of Love, Betrayal, and Quantum Physics.*

I believe that scientific theories are a means of going—somewhat mysteriously—beyond what we observe of the physical world, of penetrating into the structure of nature. The theoretical parts of scientific theories—the parts that speak in nonobservational terms—aren't, I believe, ultimately translatable into observations, but neither are they just algorithmic black boxes into which we feed our observations and churn out our predictions. I believe that the theoretical parts of theories have descriptive content and are true (or false) in the same prosaic way that the observational parts are true (or false). They're true if (and only if) they correspond to reality.

That the penetration into unobservable nature is accomplished via abstract mathematics is a large part of what makes it mystifying—mystifying enough to be coherently (if unpersuasively, at least to me) denied by scientific antirealists. It's difficult to explain exactly how science manages to do what it does—

notoriously difficult when you are trying to explain how quantum mechanics, in particular, describes unobserved reality. The unobservable aspects of nature that yield themselves to our knowledge must be both mathematically expressible and connected to our observations in requisite ways. The seventeenth-century titans, men like Galileo and Newton, figured out how to do this—how to wed mathematics to empiricism. It wasn't obvious that it was going to work, or even get us farther into nature's secrets than the Aristotelian teleological methodology it was supplanting. They made a lot of assumptions about the mathematical nature of the world and its fundamental correspondence to our cognitive modes (a correspondence they saw as reflective of God's friendly intentions toward us) in order to justify their methodology.

I also believe that since not all the properties of nature are mathematically expressible (Why should they be? It takes a very special sort of property to be expressible in that way), there are aspects of nature we will never get to by way of science; thus our scientific theories—just like our formalized mathematical systems (as proved by Gödel)—must be forever incomplete. The very fact of consciousness—an aspect of the material world we know about but not because science revealed it to us—demonstrates this necessary incompleteness.

Stuart A. Kauffman

STUART A. KAUFFMAN is external professor at the Santa Fe Institute and research professor of cell biology and physiology at the University of New Mexico. He is the author of *The Origins of Order* and *Investigations*.

Is there a fourth law of thermodynamics, or some cousin of it, concerning self-constructing nonequilibrium systems, such as biospheres, anywhere in the cosmos?

I like to think there may be such a law.

Consider this: The number of possible proteins 200 amino acids long is 20 raised to the 200th power, or about 10^{260}. The number of particles in the known universe is about 10^{80}. Suppose that on a microsecond time scale the universe were doing nothing but producing proteins 200 amino acids long. It turns out that it would take vastly many repeats of the history of the universe to create all the possible proteins of that length. For entities of a complexity greater than that of atoms—like such modestly complex organic molecules as proteins (let alone species, automobiles, or operas)—the universe is on a unique trajectory (ignoring quantum mechanics for the moment). At modest levels of complexity and above, the university is hugely non-ergodic; that is, it does not repeat itself.

Now consider the "adjacent possible," the set of entities one step away from what exists now. For chemical-reaction systems, the adjacent possible from a set of actual (already existing) compounds is the set of new compounds that can be produced by single chemical reactions among the actual set. Earth's biosphere has been expanding into its molecular adjacent possible for some 4 billion years.

Before life, there were perhaps a few hundred organic-molecule species on Earth; now there are perhaps a trillion or more. We have no law governing this expansion into the adjacent possible in this non-ergodic process. My hoped-for law is that biospheres everywhere in the universe expand as fast as possible while maintaining the rough diversity of what already exists. The law otherwise stated: The diversity of things that can happen next increases, on average, as fast as it can.

Leonard Susskind

LEONARD SUSSKIND is the Felix Bloch Professor of Theoretical Physics at Stanford University. He is the coauthor (with James Lindesay) of *An Introduction to Black Holes, Information and the String Theory Revolution: The Holographic Universe.*

[Conversation with a Slow Student:]

STUDENT: Hi, Prof. I've got a problem. I decided to do a little probability experiment—you know, coin flipping—and check some of the stuff you taught us. But it didn't work.

PROFESSOR: Well, I'm glad to hear you're interested. What did you do?

STUDENT: I flipped this coin 1,000 times. You remember, you taught us that the probability of getting heads is one-half. I figured that meant that if I flip 1,000 times, I ought to get 500 heads. But it didn't work. I got 513. What's wrong?

PROFESSOR: Yeah, but you forgot about the margin of error. If you flip a certain number of times, the margin of error is about the square root of the number of flips. For 1,000 flips, the margin of error is about 30. So you were within the margin of error.

STUDENT: Ah, now I get it! Every time I flip 1,000 times, I will always get something between 470 and 530 heads. Every single time! Wow, now that's a fact I can count on!

PROFESSOR: No, no! What it means is that you will *probably* get between 470 and 530.

STUDENT: You mean I could get 200 heads? Or 850 heads? Or even all heads?

PROFESSOR: Probably not.

STUDENT: Maybe the problem is that I didn't make enough flips. Should I go home and try it a million times? Will that work better?

PROFESSOR: Probably.

STUDENT: Aw, come on, Prof. Tell me something I can trust. You keep telling me what "probably" means by giving me more "probably"s. Tell me what probability means without using the word "probably."

PROFESSOR: Hmmm. Well how about this: It means I would be surprised if the answer were outside the margin of error.

STUDENT: My god! You mean all that stuff you taught us about statistical mechanics and quantum mechanics and mathematical probability—all it means is that you'd personally be surprised if it didn't work?

PROFESSOR: Well, uh. . . .

If I were to flip a coin a million times I'd be damn sure I wasn't going to get all heads. I'm not a betting man, but I'd be so sure that I'd bet my life or my soul on it. I'd even go the whole way and bet a year's salary. I'm absolutely certain that the laws of large numbers—probability theory—will work and protect me. All of science is based on it. But I can't prove it, and I don't really know why it works. That may be the reason Einstein said, "God does not play dice." It probably is.

Donald D. Hoffman

DONALD D. HOFFMAN is a professor of cognitive science, philosophy, and information and computer science at the University of California, Irvine; he is the author of *Visual Intelligence: How We Create What We See*.

I believe that consciousness and its contents are all that exists. Spacetime, matter, and fields never were the fundamental denizens of the universe but have always been among the humbler contents of consciousness, dependent on it for their very being.

The world of our daily experience—the world of tables, chairs, stars, and people, with their attendant shapes, smells, feels, and sounds—is a species-specific user interface between ourselves and a realm far more complex, whose essential character is conscious.

It is unlikely that the contents of our interface in any way resemble that realm; indeed, the usefulness of an interface requires, in general, that they do not. The point of an interface (such as the Windows interface on a computer) is simplification and ease of use. We click on icons because it's quicker and less error-prone than editing megabytes of software or toggling voltages in circuits. Evolutionary pressures dictate that our species-specific interface—this world of our daily experience—should itself be a radical simplification, selected not for the exhaustive depiction of truth but for the mutable pragmatics of survival.

If this is right—if consciousness is fundamental—then we should not be surprised that despite centuries of effort by the most brilliant minds there is as yet no physicalist theory of consciousness—no theory that explains how mindless matter or energy or fields could be, or cause, conscious experience. There are many proposals for where to find such a theory—perhaps in information theory, complexity, neurobiology, neural Darwinism, discriminative mechanisms, quantum effects, or functional organization. But no proposal remotely approaches the minimal standards for a scientific theory: quantitative precision and novel prediction. If matter is one of the humbler products of consciousness, then we should not expect consciousness to be theoretically derived from matter.

The mind-body problem will be to physicalist ontology what blackbody radiation was to classical mechanics: first a goad to its heroic defense, later the provenance of its final supersession. The heroic defense of physicalist ontology will, I suspect, not soon be abandoned, for the defenders doubt that a replacement grounded in consciousness could attain the mathematical precision or impressive scope of physicalist science. It remains to be seen to what extent and how effectively mathematics can model consciousness. But there are fascinating hints: According to some of its interpretations, the mathematics of quantum theory is already a major advance in this project, and perhaps much of the mathematical progress in the perceptual and cognitive sciences can also be so interpreted. We shall see.

The mind-body problem may not fall within the scope of physicalist science, since this problem has as yet no bona-fide physicalist theory. Its defenders can argue that this means only that we have not been clever enough—or that until the right mutation comes along, we cannot be clever enough—to devise a physicalist theory. They may be right. But if we assume that consciousness is fundamental, then the mind-body problem changes from an attempt to bootstrap consciousness from matter into an attempt to bootstrap matter from consciousness. The latter bootstrap is, in principle, elementary: Matter, fields, and spacetime are among the contents of consciousness.

The rules by which, for instance, human vision constructs colors, shapes, depths, motions, textures, and objects—rules now emerging from psychophysical and computational studies in the cognitive sciences—can be read as a description, partial but mathematically precise, of this bootstrap. What we lose in this process are physical objects that exist independent of any observer. There is no sun or moon, unless a conscious mind perceives them; both are constructs of consciousness, icons in a species-specific user interface. To some this seems a *reductio ad absurdum* readily contradicted by experience and our best science. But our *best* science, which is our theory of the quantum, gives no such assurance, and experience once led us to believe that the earth was flat and the stars were near. Perhaps mind-independent objects will one day go the way of the flat earth.

This view obviates no methods or results of science but integrates and reinterprets them in its framework. Consider for instance the quest for the neural correlates of consciousness. This holy grail of physicalism can and should proceed, if consciousness is fundamental, for it constitutes a central investigation of our user interface. To the physicalist, such neural correlates are potentially a causal source of consciousness. But if consciousness is fundamental, then its neural correlates are a feature of our interface, corresponding to, but never causally responsible for, alterations of consciousness. Damage the brain, destroy the neural correlates, and consciousness is, no doubt, impaired. Yet neither the brain nor the neural correlates cause consciousness; instead, consciousness constructs the brain. This is no mystery. Drag a file's icon to the recycle bin and the file is, no doubt, deleted. Yet neither the icon nor the recycle bin, each a mere pattern of pixels on a screen, causes its deletion. The icon is a simplification, a graphical correlate of the file's contents, intended to hide, not to instantiate, the complex web of causal relations.

Terrence Sejnowski

TERRENCE SEJNOWSKI, a computational neuroscientist, is an investigator with the Howard Hughes Medical Institute and divides his time between the Salk Institute for Biological Studies and the University of California at San Diego, where he investigates the principles linking brain mechanisms and behavior. He is coauthor (with Patricia Churchland) of *The Computational Brain*.

How do we remember the past?

There are many answers to this question, depending on whether you are an artist, a historian, or a scientist. As a scientist, I want to know the mechanisms reponsible for storing memories and where in the brain memories are stored. Although neuroscientists have made tremendous progress in uncovering neural mechanisms for learning, I believe (but cannot yet prove) that we are all looking in the wrong place for where long-term memories are stored.

I have been puzzled by my ability to remember my childhood even though most of the molecules in my body today are not the same ones I had as a child—in particular, the molecules that make up my brain are constantly being replaced with newly minted molecules. Despite this molecular turnover, I have detailed memories of places where I lived fifty years ago—memories that I never rehearsed but which are easily verified.

If memories are stored as changes to molecules inside brain cells—molecules that are constantly being replaced—how can a memory remain stable over fifty years? My hunch is that the substrate of old memories is located not inside the cells but outside, in the extracellular space. That space is not empty but filled with a matrix of tough material that connects cells and helps them maintain their shape. Like scar tissue, the matrix is difficult to dissolve and is replaced very slowly, if at all. (This explains why scars on your body haven't changed much after decades of sloughing off skin cells.)

My intuition is based on a set of classic experiments on the junction between motor neurons and muscle cells. When the neuromuscular junction is activated, the muscle contracts. If the nerve that activates a muscle is crushed, the nerve fiber grows back to the junction, forming a specialized nerve terminal ending. This occurs even if the muscle cell is also killed. The "memory" of the contact in this case is preserved by the extracellular matrix at the neuromuscular junction, called the basal lamina. The extracellular matrix at synapses in the brain may have a similar function and could well maintain overall connectivity despite the comings and goings of molecules inside neurons.

How could we prove that the extracellular matrix is responsible for long-term memories? The theory predicts that if the extracellular matrix is disrupted, memories will be lost. This experiment can be done with enzymes that selectively degrade components of the extracellular matrix or by knocking out one or more key molecules using molecular genetic techniques. If I'm right, then all of your memories—what makes you a unique individual—are contained in the brain's exoskeleton.

The intracellular machinery holds memories temporarily and decides what to permanently store in the extracellular matrix, perhaps while you are sleeping. It might be possible someday to stain this memory exoskeleton and see what our memories look like.

John Horgan

JOHN HORGAN is a freelance science journalist and author. He is the author of several books, including *The End of Science* and *Rational Mysticism: Dispatches from the Border Between Science and Spirituality*.

I believe that neuroscientists will never have enough understanding of the neural code, the secret language of the brain, to read peoples' thoughts without their consent.

The neural code is the software, algorithm, or set of rules whereby the brain transforms raw sensory data into perceptions, memories, decisions, meanings. A complete solution to the neural code could, in principle, allow scientists to monitor and manipulate minds with exquisite precision. You might, for example, probe the mind of a suspected terrorist for memories of past attacks or plans for future ones. The problem is that although all brains operate according to certain general principles, each person's neural code is idiosyncratic, shaped by his or her unique life history.

The neural pattern that underpins my concept of "George Bush" or "Heathrow Airport" or "surface-to-air missile" differs from yours. The only way to know how my brain encodes this kind of specific information would be to monitor its activity—ideally with thousands or even millions of implanted electrodes that can detect the chatter of individual neurons—while I tell you as precisely as possible what I am thinking. But the data that you glean from studying me will be of no use for interpreting the signals of any other person. For ill or good, our minds will always remain hidden to some extent from Big Brother.

Arnold Trehub

ARNOLD TREHUB is adjunct professor of psychology at the University of Massachusetts, Amherst. He has been the director of a laboratory devoted to psychological and neurophysiological research and is the author of *The Cognitive Brain*.

I have proposed a law of conscious content, which asserts that for any experience, thought, question, or solution there is an analog in the biophysical state of the brain. As a corollary to this principle, I have argued that the conventional attempts to understand consciousness simply by searching for its neural correlates (in both theoretical and empirical investigations) are too weak to provide a good understanding of conscious content. Instead, I have proposed that we go beyond this and explore brain events that have at least some similarity to our phenomenal experiences—namely, neuronal *analogs* of conscious content. In support of this approach, I have presented a theoretical model that does more than address the sheer correlation between mental states and neuronal events in the brain. It explains how neuronal analogs of phenomenal experience can be generated, and it details how essential human cognitive tasks can be accomplished by the particular structure and dynamics of putative neuronal mechanisms and systems in the brain.

A large body of experimental findings, clinical findings, and phenomenal reports can be explained within a coherent framework by the neuronal structure and dynamics of my theoretical model. In addition, the model accurately predicts many classical illusions and perceptual anomalies. So I believe that the neuronal mechanisms and systems I have proposed provide a true explanation for many important aspects of human cognition and phenomenal experience. But I can't prove it. Of course, competing theories about the brain, cognition, and consciousness can't be proved either. Providing the evidence is the best we can do—I think.

Ned Block

NED BLOCK is a professor of philosophy and psychology at New York University. He is coeditor (with Owen Flanagan and Güven Güzeldere) of *The Nature of Consciousness*.

I am optimistic that the so-called Hard Problem of consciousness will be solved by empirical and conceptual advances—working in tandem—made in cognitive neuroscience. What is the Hard Problem? No one has a clue (at the moment) how to answer the question of why the neural basis of the phenomenal feel of my experience of, for instance, red is the neural basis of that particular phenomenal feel rather than a different one or none at all. There is an explanatory gap here that we do not know how to close now, but I have faith that we will one day. The Hard Problem is conceptually and explanatorily prior to the issue of what the nature of the self is, as can be seen in part by noting that the problem would persist even for experiences that are not organized into selves. No doubt solving the Hard Problem (i.e., closing the explanatory gap) will require ideas we cannot now anticipate. The mind-body problem is so singular that no appeal to the closing of past explanatory gaps justifies optimism. But I am optimistic nonetheless.

Janna Levin

JANNA LEVIN is a theoretical physicist and a professor of physics and astronomy at Barnard College of Columbia University. She is the author of *How the Universe Got Its Spots: Diary of a Finite Time in a Finite Space*.

I believe that there is an external reality, and that you are not all figments of my imagination. My friend asks me, through the steam he blows off the surface of his coffee, how I can trust the laws of physics back to the origins of the universe. I ask him how he can trust the laws of physics down to his cup of coffee. He shows every confidence that the scalding liquid will not spontaneously defy gravity and fly up into his eyes. He lives with this confidence, born of his empirical experience of the world. His experiments with gravity, heat, and light began in childhood, when he palpated the world to test its materials. Now he has a refined and well-developed theory of physics, whether expressed in equations or not.

I simultaneously believe more and less than he does. It is rational to believe what all of my empirical and logical tests of the world confirm—that there is a reality that exists independent of me. That the coffee will not fly upward. But it is a belief nonetheless. Once I've gone that far, why stop at the perimeter of mundane experience? Just as we can test the temperature of a hot beverage with a tongue or a thermometer, we can test the temperature of the primordial light left over from the Big Bang. One is no less real than the other simply because it is remarkable. How can I know that mathematics and the laws of physics can be reasoned down to the moment of creation of time, space, the entire universe? In the very same way that my friend believes in the reality of the second double cappuccino he orders. In formulating our beliefs, we are honest and critical and able to admit when we are wrong—and these attitudes are the cornerstones of truth.

But how do I really know? If I measure the temperature of boiling water, all I really know is that mercury climbs a glass tube. Not even that: All I really know is that I see mercury climb a glass tube. But maybe the image in my mind's eye isn't real. Maybe nothing is real—not the mercury, not the glass, not the coffee, not my friend. They are all products of a florid imagination. There is no external reality, just me. Einstein? My creation. Picasso? My mind's forgery. But this solipsism is ugly and arrogant.

When I leave the café, I believe the room of couches and tables is still there, that it is still full of people, that they haven't evaporated now that I can no longer see them. But if I am wrong and there is no external reality, then not only is this essay my invention but so is the Web, edge.org, all of its participants and their ingenious ideas. And if you are reading this, I have created you, too. And if I am wrong and there is no external reality, then maybe it is me who is a figment of your imagination and the cosmos outside your door is your magnificent creation.

Daniel Gilbert

DANIEL GILBERT is the Harvard College Professor in the Department of Psychology at Harvard University and director of its laboratory on social cognition and emotion.

In the not too distant future, we will be able to construct artificial systems that give every appearance of consciousness—systems that act like us in every way. These systems will talk, walk, wink, lie, and appear distressed by close elections. They will swear up and down that they are conscious and they will demand their civil rights. But we will have no way to know whether their behavior is more than a clever trick—more than the pecking of a pigeon that has been trained to type "I am, I am!"

We take each other's consciousness on faith, because we must, but after 2,000 years of worrying about this issue, no one has ever devised a definitive test of its existence. Most cognitive scientists believe that consciousness is a phenomenon that emerges from the complex interaction of decidedly nonconscious parts (neurons), but even when we finally understand the nature of that complex interaction we still won't be able to prove that it produces the phenomenon in question. And yet I haven't the slightest doubt that everyone I know has an inner life—a subjective experience, a sense of self—that is very much like mine.

What do I believe is true but cannot prove? The answer is: You!

Todd E. Feinberg, M.D.

TODD E. FEINBERG, M.D., is professor of clinical neurology and psychiatry at the Albert Einstein College of Medicine and chief of the Betty and Morton Yarmon Division of Neurobehavior and Alzheimer's Disease at the Beth Israel Medical Center in New York. He is the author of *Altered Egos: How the Brain Creates the Self.*

I believe that the human race will never decide that an advanced computer possesses consciousness. Only in science fiction will a person be charged with murder if they unplug a PC. I believe this because I hold, but cannot yet prove, that in order for an entity to be conscious and possess a mind, it has to be a living being.

Being alive, of course, does not guarantee the presence of a mind. A plant carries on the metabolic functions of life but does not possess a mind. A chimpanzee, on the other hand, is a different story. All the behavioral features we share with chimps in addition to life—intelligence, the ability to deceive, mirror self-recognition, some individual social identity—make chimps seem so much like us that many in the scientific community intuitively grant chimps "beinghood" and consciousness.

In addition to being alive, therefore, it appears that a living thing must be a being—must possess a self—to possess a mind.

But silicon chips are not alive and computers are not beings. I argue that this is so because the particular material substance and arrangement of the brain is essential to the creation of consciousness and "beinghood." Computers will never achieve consciousness because in order to be conscious "like us," it will need to be made of living stuff like us, to grow like us, and (unfortunately) be able to die like us.

Clifford Pickover

THE COMPUTER SCIENTIST and science writer Clifford Pickover is a staff member at the IBM T. J. Watson Research Center. He is the author of numerous books, including *Sex, Drugs, Einstein, and Elves: Sushi, Psychedelics, Parallel Universes, and the Quest for Transcendence.*

If we believe that consciousness is the result of patterns of neurons in the brain, our thoughts, emotions, and memories could be replicated in moving Tinkertoy assemblies. The Tinkertoy minds would have to be very big to represent the complexity of our minds, but it nevertheless could be done, in the same way that people have made computers out of 10,000 Tinkertoy pieces. In principle, our minds could be hypostatized in patterns of twigs, in the movements of leaves, in the flocking of birds. The philosopher and mathematician Gottfried Leibniz liked to imagine a machine capable of conscious experiences and perceptions. He said that even if this machine were as big as a mill and we could explore inside, we would find "nothing but pieces which push one against the other and never anything to account for a perception."

If our thoughts and consciousness do not depend on the actual substances in our brains but rather on the structures, patterns, and relationships between parts, then Tinkertoy minds could think. If you could make a copy of your brain with the same structure but using different materials, the copy would think it was you. This seemingly materialistic approach to mind does not diminish the hope of an afterlife, of transcendence, of communion with entities from parallel universes, or of God himself. Even Tinkertoy minds can dream, seek salvation and bliss—and pray.

Nicholas Humphrey

THE THEORETICAL PSYCHOLOGIST Nicholas Humphrey is School Professor at the London School of Economics. His books include *Consciousness Regained*, *A History of the Mind*, and *The Mind Made Flesh*.

I believe that human consciousness is a conjuring trick, designed to fool us into thinking we are in the presence of an inexplicable mystery. Who is the conjuror, and what can be the point of such deception? The conjuror is the human mind itself, evolved by natural selection, and the point has been to bolster human self-confidence and self-importance—so as to increase the value we each place on our own and others' lives.

If this is right, it provides a simple explanation for why we, as scientists or laymen, find the "hard problem" of consciousness just so hard. Natural selection has meant it to be hard. Indeed, "mysterian" philosophers, from Colin McGinn to the late Pope John Paul II, who bow down before the apparent miracle and declare that it is impossible in principle to understand how consciousness could arise in a material brain, are responding exactly as natural selection hoped they would—with shock and awe.

Can I prove it? It's difficult to prove any adaptationist account of why humans experience things the way they do. And in this case there may be an added catch. For just to the extent that natural selection has succeeded in putting consciousness beyond the reach of rational explanation, it will have undermined the very possibility of *showing* that this is what it has done.

Nonetheless there may be a loophole by which science could still enter. While it might seem—and even be—impossible to explain how a brain process could actually *have* the quality of consciousness, it might not be at all impossible to explain how a brain process could be designed to give rise to the *impression* of having this quality. Consider: We could never explain why 2 + 2 = 5, but we might relatively easily be able to explain why someone should be under the illusion that 2 + 2 = 5.

Would I want to explain consciousness this way, if I could? That's a difficult one. If the illusion that consciousness is an inexplicable mystery is a source of human hope, I suppose there is a real danger that exposing the trick might send us all to Hell.

Pamela McCorduck

THE WRITER AND HISTORIAN Pamela McCorduck has written several books on artificial intelligence and the intellectual impact of computers, including *Machines Who Think*.

Although I can't prove it, I believe that thanks to new kinds of social modeling that take into account individual motives as well as group goals, we will soon grasp in a deep way how collective human behavior works, whether it's action by small groups or by nations. Any predictive power this understanding has will be useful, especially with regard to unexpected outcomes and even unintended consequences. But it will not be infallible, because the complexity of such behavior makes exact prediction impossible.

Charles Simonyi

THE COMPUTER SCIENTIST Charles Simonyi worked for the Xerox Corporation's Palo Alto Research Center (PARC) from 1972 to 1980 and then joined Microsoft to start the development of microcomputer application programs. He hired and managed teams who developed Microsoft Multiplan, Word, Excel, and other applications. In 2002 he founded Intentional Software Corporation, a company dedicated to perfecting software relationships.

I believe we are writing software the wrong way. There are sound evolutionary reasons for why we are doing so, which we can call the programming-the-problem-in-a-computer-language paradigm, but the incredible success of Moore's law (that computer capacity doubles approximately every eighteen months) blinded us to the fact that we are stuck in an evolutionary backwater. Computers are demonstrably 10,000 times better than they were not so long ago, yet we are not seeing their services improve at the same rate (with some exceptions—for example, games and Internet searches). An administrative problem, say, that would take maybe 100 pages to describe precisely will take millions of dollars to program for a computer, and often the program will not work.

Recently a small airline came to a standstill because of a

problem in crew-scheduling software—raising the ire of Congress, not to mention that of their customers. My laptop could store 200 pages of text (1/2 megabytes) for every crew member of this airline just in its fast memory and 100 times more (a veritable encyclopedia of 20,000 pages) for each person on its hard disk. But for a schedule we would need only one or two—at most ten—pages per crew member. Even with all the rules—the laws; the union contracts; the local, state, federal taxes; the duty-time limitations; the FAA regulations on crew certification—is there anyone who thinks the problem is not simple in terms of computing? We need to store and process, at the maximum, ten pages per person, when we have capacity for 2,000 times more than that in one cheap laptop! Of course, the problem is complex in terms of the problem domain, but not shockingly so. All the rules relevant to aircraft-crew scheduling are probably expressible in less than 1,000 pages—or 0.5 percent of the fast memory.

Software is the bottleneck on the high-tech horn of plenty. The scheduling program for the airline takes up far more memory than it should; hence, the software represents complexity far greater than that of the problem itself. No wonder some planes are assigned three pilots by the software, while the others can't fly because the copilot is not scheduled. Note that the cost of memory is not the issue—we could afford that waste. But the use of so much memory for software is an indication of complexity inflation occurring during programming.

What is going on? I like to use cryptography as the metaphor. We take a message and combine it with a key, using a difficult-to-invert function to get the code. Programmers using today's paradigm start from a problem statement—for example, that a Boeing 767 needs a pilot, a copilot, and seven cabin crew, with various certification requirements for each—and combine this with their knowledge of computer science and software engineering. That is how this rule can be encoded in computer language and turned into an algorithm. The act of combining is the programming process, whose result is the source code. Now, programming is well known to be a difficult-to-invert function, though perhaps not to cryptography's standards. But one can joke that the airline could keep its proprietary scheduling rules secret by publishing the source code for their implementation, since no one could ever figure it out—or even whether the published code had to do with scheduling or spare-parts inventory. It can be that obscure.

The amazing thing is that today it is the source code—i.e., the encrypted problem—that software engineering focuses on. To add insult to injury, the "encryption"—i.e., the programming—is done manually, which means high costs, low speed, and high error rates. When a general in the field realizes that the message he wants to send his lieutenants contains incorrect information, he would not think of having the message edited after its encryption (or of "fixing the code"); instead, the original text is edited, and then this improved message is re-encrypted. The message may be wrong, but it won't be wrong because of the encryption, and it is easily fixed.

The airline problem statement above is obviously oversimplified. But viewed through the funhouse mirror of software coding, it becomes all but unrecognizable: 1,000 times fatter, disjointed, foreign. What can be done? Follow the metaphor. First, focus on recording the problem statement—the "original message" in our battlefield cryptography metaphor. This is not a program in any sense of the word, just a straightforward recording of the subject matter experts' concerns, using their own jargon, their own notations. Next, empower the programmers to write not a program for the problem itself but a program generator, which will combine the subject matter experts' contributions with implementation detail and output the code that the programmers would have written as a direct solution. This is called generative programming. The generator is a mechanized expression of the programmers' expertise, and it effectively separates issues of subject matter from issues of software engineering, so that most changes on either side do not involve the other—since the other side's repetitive contribution is expressed either by the generator or by the recorded problem statement. For this reason, I believe that generative programming is the future of software.

Alan Kay

THE COMPUTER SCIENTIST Alan Kay joined Xerox Corporation's Palo Alto Research Center (PARC) in 1970 and was one of the key members to develop prototypes of networked workstations using the programming language Smalltalk. He is also one of the fathers of the idea of object-oriented programming and the conceiver of the Dynabook concept, which defined the basics of the laptop. He is currently president of Viewpoints Research Institute and a senior fellow at Hewlett Packard Laboratories.

Einstein said "You must learn to distinguish between what is true and what is real." Science is a relationship between what we can represent and think about and what's actually "out there;" it's an extension of good mapmaking. When we guess in science, we are guessing about approximations and mappings to languages, not guessing about "the truth"—and we are not in a good state of mind for doing science if we think we are "guessing the truth" or "finding the truth." This is not at all well understood outside science, and unfortunately some people with science degrees don't seem to understand it either.

There are, for example, very few interesting actual proofs in computing. (The eminent programmer Don Knuth of Stanford likes to say, "Beware of bugs in the above code; I have only proved it correct, not tried it.") We'd like to prove useful programs correct, but we either have intractable degrees of freedom or (as in the Knuth quote) it's very difficult to know whether we've taken all cases into account. So a guess in computing is often architectural or a collection of covering heuristics.

One guess I made long ago—and which does not yet have a body of evidence to support it—is that what's special about the computer is analogous to (and an advance on) what was special about writing and then printing. It's not the automating of past forms that has the impact. As the media philosopher Marshall McLuhan pointed out, when you change the nature of representation and argumentation, people who learn in these new ways will turn out to be qualitatively different thinkers (and better thinkers?), and this will (usually) advance our limited conceptions of civilization.

This still seems like a good guess to me—but "truth" has nothing to do with it.

Steven Pinker

STEVEN PINKER, an experimental psychologist, is Johnstone Family Professor in the Department of Psychology at Harvard University and the author of, among other books, *The Language Instinct*, *How the Mind Works*, *Words and Rules*, and *The Blank Slate: The Modern Denial of Human Nature*.

In 1974 Marvin Minsky wrote that "there is room in the anatomy and genetics of the brain for much more mechanism than anyone today is prepared to propose." Today, many advocates of evolutionary and domain-specific psychology are willing to propose the richness of mechanism that Minsky called for thirty years ago. For example, I believe that the mind is organized into cognitive systems specialized for reasoning about objects, space, numbers, living things, and other minds; that we are equipped with emotions triggered by other people (sympathy, guilt, anger, gratitude) and by the physical world (fear, disgust, awe); that we have different ways for thinking and feeling about people in different kinds of relationships to us (parents, siblings, other kin, friends, spouses, lovers, allies, rivals, enemies); and several peripheral drivers for communicating with others (language, gesture, facial expression).

When I say I believe this but cannot prove it, I don't mean

that it's a matter of raw faith or even an idiosyncratic hunch. In each case, I can provide reasons for my belief, both empirical and theoretical. But I certainly can't prove it, or even demonstrate it in the way that molecular biologists demonstrate their claims—namely, in a form so persuasive that skeptics can't reasonably attack it and a consensus is rapidly achieved. The idea of a richly endowed human nature is still unpersuasive to many reasonable people, who often point to certain aspects of neuroanatomy, genetics, and evolution that appear to speak against it. I believe, but cannot prove, that these objections will be met as the sciences progress.

From the standpoint of neuroanatomy and neurophysiology, critics have pointed to the apparent homogeneity of the cerebral cortex and the seeming interchangeability of cortical tissue in experiments in which patches of cortex are rewired or transplanted in animals. I believe that the homogeneity is an illusion, owing to the fact that the brain is a system for information processing. Just as all books look the same to someone who does not understand the language in which they are written, and the DVDs of all movies look the same under a microscope, the cortex may look homogeneous to the eye but nonetheless contain different patterns of connectivity and synaptic biases that allow it to compute very different functions. I believe that these differences will be revealed in different patterns of gene expression in the developing cortex. I also believe that the apparent interchangeability of cortex occurs only in early stages of sensory systems that happen to have similar computational demands, such as isolating sharp signal transitions in time and space.

From the standpoint of genetics, critics have pointed to the small number of genes in the human genome (now thought to be less than 25,000) and to their similarity to those of other animals. I believe that geneticists will find that there is a large store of information in the noncoding regions of the genome (the so-called junk DNA), whose size, spacing, and composition could have large effects on how genes are expressed. The genes themselves may code largely for the meat and juices of the organism, which are pretty much the same across species, whereas how gene products are sculpted into brain circuits may depend on a much larger body of genetic information. I also believe that many examples of what we call "the same genes" in different species may differ in tiny ways at the sequence level—ways that have large consequences for how the organism is put together.

And at the level of evolution, critics have pointed to the difficulty in establishing the adaptive function of a psychological trait. I believe this difficulty will vanish as we come to understand the genetic basis of psychological traits in more detail. New techniques in genomic analysis, which look for statistical fingerprints of selection in the genome, will show that many genes involved in cognition and emotion were specifically selected for in the primate, and in many cases the human, lineage.

Christine Finn

CHRISTINE FINN is an archaeologist and journalist based in Rome and Visiting Fellow in the Department of Archaeology and Anthropology at the University of Bristol, U.K. She is the author of *Past Poetic* and *Artifacts: An Archaeologist's Year in Silicon Valley.*

I believe that modern humans greatly underutilize their cognitive capabilities. Proving this, however, would mean embracing the very same sentient possibilities—visceral hunches—that were possibly part of the world of archaic humans. This enlarged realm of the senses acknowledges reason but also heeds the grip of the gut, the body poetic.

Daniel C. Dennett

DANIEL C. DENNETT is University Professor and Austin B. Fletcher Professor of Philosophy and director of the Center for Cognitive Studies at Tufts University. He is the author of, among other books, *Consciousness Explained, Darwin's Dangerous Idea*, and *Freedom Evolves*.

I believe, but cannot yet prove, that acquiring a human language (an oral or sign language) is a necessary precondition for consciousness—in the strong sense of there being a subject, an I, a "something it is like something to be." It would follow that nonhuman animals and prelinguistic children—although they can be sensitive, alert, responsive to pain and suffering, and cognitively competent in many remarkable ways (including ways that exceed normal adult human competence)—are not really conscious, in this strong sense: There is no organized subject (yet) to be the enjoyer or sufferer, no owner of the experiences as contrasted with a mere cerebral locus of effects.

This assertion is shocking to many people, who fear that it would demote animals and prelinguistic children from moral protection, but this would not follow. Whose pain is the pain occurring in the newborn infant? There is not yet anybody whose pain it is, but that fact would not license us to inflict painful stimuli on babies or animals, any more than we are licensed to abuse the living bodies of people in comas, who are definitely not conscious. If selfhood develops gradually, then certain types of events only gradually become experiences, and there will be no sharp line between unconscious pains (if we may call them that) and conscious pains, and both will merit moral attention. (And, of course, the truth of the empirical hypothesis is in any case strictly independent of its ethical implications, whatever they are. Those who shun the hypothesis on purely moral grounds are letting wishful thinking overrule a properly inquisitive scientific attitude. I am happy to give animals and small children the benefit of the doubt for moral purposes, but not for scientific purposes.) Those who are shocked by my hypothesis should pause, if they can bear it, to notice that it is as just as difficult to prove its denial as its assertion. But it can, I think, be proved eventually. Here's what it will take, one way or the other:

1. a well-confirmed model of the functional architecture of adult human consciousness, showing how long-distance pathways of reverberant interactions in the cortex have to be laid down and sustained by the sorts of cascades of self-stimulation children engage in when they are first acquiring language;

2. an interpretation of the dynamics of the model that explains why, absent these well-traveled pathways of neural micro habit, there is no functional unity to the nervous system—no unity to distinguish an I from a we (or from a multitude) as the candidate subject(s) occupying that nervous system;

3. further experimental work demonstrating the importance of what Thomas Metzinger calls "the phenomenal model of the intentionality relation" in enabling the sorts of experiences we consider central to our own adult consciousness. This work will

demonstrate that animal cleverness never requires the abilities thus identified in humans, and that animals are incapable of appreciating many things we normally take for granted as aspects of our conscious experience.

This is an empirical hypothesis, and it could just as well be proved false. It could be proved false by showing that the necessary pathways functionally uniting the relevant brain systems (in the ways I claim are required for consciousness) are already provided in normal infant or fetal development and are present in, say, all mammalian nervous systems of a certain maturity. I doubt this is true, because it seems clear to me that evolution has already demonstrated that remarkable varieties of adaptive coordination can be accomplished without such hyper-unifying meta-systems—by colonies of social insects, for instance. What is it like to be an ant colony? Nothing, I submit, and I think most people would agree intuitively. What is it like to be a brace of oxen? Nothing (even if it is like something to be a single ox). But then we have to take seriously the extent to which animals—not just insect colonies and reptiles but rabbits, whales, bats, and chimpanzees—can get by with somewhat disunified brains.

Evolution will not have provided for the further abilities where they were not necessary for members of those species to accomplish the tasks their lives pose them. If animals were like the imaginary creatures in the fictions of Beatrix Potter or Walt Disney, they would have to be conscious pretty much the way we are. But animals are more different from us than we usually imagine, enticed as we are by those charming anthropomorphic fictions. We need these abilities to become persons, communicating individuals capable of asking and answering, requesting and forbidding and promising (and lying). But we don't need to be born with those abilities, since normal rearing will entrain the requisite neural dispositions. Human subjectivity, I am proposing, is thus a remarkable by-product of human language, and no version of it should be extrapolated to any other species by default, any more than we should assume that the rudimentary communication systems of other species have verbs and nouns, prepositions and tenses.

Finally, since there is often misunderstanding on this score, I am not saying that all human consciousness consists in talking to oneself silently, although a great deal of it does. I am saying that the ability to talk to yourself silently, as it develops, also brings along with it the abilities to review, to muse, to rehearse, to recollect, and in general to engage the contents of events in one's nervous system that would otherwise leave no memories in their wake and hence contribute to one's guidance in ways that are well described as unconscious. If a nervous system can come to sustain all those abilities without having language, then I am wrong.

Alun Anderson

ALUN ANDERSON was the editor-in-chief of *New Scientist* from 1992 to 2005 and is currently senior consultant to the magazine.

I believe that cockroaches are conscious. That is probably an unappealing thought to anyone who switches on a kitchen light in the middle of the night and finds a roach family running for cover. But it's really shorthand for saying I believe that many quite simple animals are conscious, including more attractive beasts, like bees and butterflies.

I can't prove they are, but I think in principle it will be provable one day, and there's a lot to be gained from thinking about the worlds of these relatively simple creatures, both intellectually and even poetically. I don't mean they are conscious in the same way humans are; if that were true, the world would be a boring place. Rather, the world is full of many overlapping alien consciousnesses.

Why do I think there might be multiple forms of consciousness out there? Before becoming a journalist, I spent ten years and a couple of postdoctoral fellowships getting inside the sensory worlds of a variety of insects, including bees and cockroaches. I was inspired by *A Stroll Through the Worlds of*

Animals and Men: A Picture Book of Invisible Worlds, a slim volume, now out of print, by Jakob von Uexküll (1864–1944).

The same book had also inspired Niko Tinbergen and Konrad Lorenz, the Nobel laureates who founded the field of ethology (animal behavior). Von Uexküll studied the phenomenal world of animals—what he called their *Umwelt*, the world around animals as they perceive it. Everything an animal senses means something to it, for it has evolved to fit and create its world. The study of animals and their sensory worlds has now morphed into the field of sensory ecology, which seeks to relate perceptual systems to an animal's lifestyle, or, on a wilder path, the newer science of biosemiotics.

I spent time studying how honeybees could find their way around my laboratory (they had learned to fly in through a small opening in the window) and find a hidden source of sugar. Bees could learn all about the pattern of key features in the room and would show that they were confused if objects had been moved while they were out of the room. They were also easily distracted by certain kinds of patterns—particularly ones with lots of points and lines having abstract similarities to the patterns on flowers—and by floral scents, and also by sudden movements that might signal danger. In contrast, when they were busy gorging on the sugar almost nothing could distract them, making it possible for me to paint a little number on their backs so that I could distinguish individual bees.

To make sense of this ever-changing behavior with its shifting focus of attention, I always found it simplest to figure out what was happening by imagining the sensory world of the bee, whose eyes are extraordinarily sensitive to flicker and to colors we can't see. I imagined it as a visual screen, in the same way that I can sit back and "see" my own visual screen of everything happening around me, with sights and sounds coming in and out of prominence. The objects in the bees' world have significances or "meanings" quite different from our own, which is why a bee's attention is drawn to things we would barely perceive.

That's what I mean by "consciousness"—the feeling of "seeing" the world and its associations. For the bee, it is the feeling of being a bee. I don't mean that a bee is self-conscious or spends time thinking about itself. But the problem of how the bee has its own "feeling" is the same incomprehensible "hard problem" of how the activity of our nervous system gives rise to our own "feelings."

At least the bee's world is highly visual and capable of being imagined. Some creatures live in sensory worlds that are much harder to access. Spiders that hunt at night live in a world dominated by the detection of faint vibration and of the tiniest flows of air that allow them to sense a passing fly in pitch darkness. The sensory hairs covering their body give them a sensitivity to touch far more fine-grained than that of our own skin.

To think this way about simple creatures is not to fall into the anthropomorphic fallacy. Bees and spiders live in their own worlds, in which I don't see humanlike motives. Rather, it is a kind of panpsychism I am quite happy to own up to—at least until we know a lot more about the origin of consciousness. This may take me out of the company of quite a few scientists who would prefer to believe that a bee with a brain containing only a million neurons must surely be a collection of instinctive reactions with some simple switching mechanism between them, rather than an entity with some central representation of what is going on which might be called consciousness. But it leaves me in the company of poets, who wonder at the world of even lowly creatures.

In this falling rain,
where are you off to
snail?
wrote the haiku poet Issa.

As for the cockroaches, they are a little more human than the spiders. Like the owners of the New York apartments who detest them, they suffer from stress and can die from it, even without injury. They are also hierarchical and they know their little territories well. When they are running for it, think twice before crushing out another world.

Joseph LeDoux

JOSEPH LEDOUX is a University Professor and Henry and Lucy Moses Professor of Science at New York University. He is the author of *The Emotional Brain* and *Synaptic Self: How Our Brains Become Who We Are*.

I believe that animals have feelings and other states of consciousness, but neither I nor anyone else has been able to prove it. We can't even prove that other people are conscious, much less other animals. In the case of other people, though, we at least can have a little confidence, since all people have brains with the same basic configurations. But as soon as we turn to other species and start asking questions about feelings, and about consciousness in general, we are in risky territory, because the hardware is different.

When a rat is in danger, it does things that many other animals do: That is, it either freezes, runs away, or attacks. People pretty much do the same. Some scientists say that because a rat and a person act the same in similar situations, they have the same kinds of subjective experience. I don't think we can say this.

There are two aspects of brain hardware that make it difficult for us to generalize from our subjective experiences to the experiences of other animals. One is that the circuits most often associated with human consciousness involve the lateral prefrontal cortex (via its role in working-memory and executive-control functions). This broad zone is much more highly developed in people than in other primates, and does not seem to exist in other creatures. So certainly for those aspects of consciousness that depend on the prefrontal cortex, including our knowledge of who we are and our ability to make plans and decisions, there is reason to believe that even other primates might differ from people. Another dramatic difference is that humans have natural language. Because so much of human experience is tied up with language, consciousness is often said to depend on it. If so, then other animals are ruled out of the consciousness game. But even if consciousness doesn't depend on language, language certainly changes consciousness, so that whatever consciousness another animal has is likely to differ from most of our states of consciousness.

For these reasons, it is hard to know what consciousness might be like in another animal. If we can't measure it (because it is internal and subjective) and can't use our own experience to frame questions about it (because the hardware that makes it possible is different), it becomes difficult to study.

Most of what I have said applies to the content of conscious experience. There is another aspect of consciousness that is less problematic scientifically. It is possible to study the processes that make consciousness possible in other animals, even if we can't study the content of their consciousness. This is exactly what is done in studies of working memory in nonhuman primates. One approach that has had some success in the area of conscious content in nonhuman primates has focused on a limited kind of consciousness, visual awareness. But this approach, by Christof Koch and Francis Crick, investigates the neural correlates of consciousness rather than the causal mechanisms. The correlates and the mechanisms may be the same—but they may not. Interestingly, this approach also emphasizes the importance of the prefrontal cortex in making visual awareness possible.

So what about feelings? My view is that a feeling is what happens when an emotion system, like the fear system, is active in a brain that can be aware of its own activities. That is, what we call "fear" is the mental state we are in when the activity of the defense system of the brain (or the consequences of its activity, such as bodily responses) occupies working memory. Viewed this way, feelings are strongly tied to those areas of the cortex that are fairly unique to primates and especially well developed in people. The addition of natural language makes for fine gradations of feeling, because it allows us to use words and grammar to differentiate and categorize states and to attribute them not just to ourselves but to others.

There are other views about feelings: Antonio Damasio argues that feelings arise from more primitive activity in body-sensing areas of the cortex and brainstem. Jaak Pankseep has a similar view, though he focuses more on the brainstem. Because this network has not changed much in the course of human evolution, it could therefore be involved in feelings that are shared across species. I don't object to this notion on theoretical grounds, but I don't think it can be proved. Pankseep argues that if it looks like fear in rats and people, it probably feels like fear in both species, but how do you know that rats and people feel the same way when they behave the same way? A cockroach will escape from danger—does it, too, feel fear as it runs away? I don't think behavioral similarity is sufficient grounds for proving that experience is similar. Neural similarity helps—rats and people have similar brainstems, and a roach doesn't even have a brain. But is the brainstem responsible for feelings? Even if that were proved to be the case in people, how would you prove it in a rat?

So we're back where we started. I think rats and other mammals, and maybe even roaches (who knows?), have feelings. But I don't know how to prove it. And because I have reason to think that their feelings might be fundamentally different from ours (since human consciousness seems to depend on special circuits and on language), I prefer to study emotional behavior in rats, rather than emotional feelings. I study rats because you can make progress at the neural level, provided that what you measure is the same in rats and people. I wouldn't study language or consciousness in rats, so I don't study feelings either, because I don't know that they exist. I may be accused of short-sightedness for this, but I'd rather make progress on something I can study in rats than beat my head against the consciousness wall in these creatures. I'm a practical emotionalist.

George Dyson

GEORGE DYSON is a historian of technology whose interests have ranged from the history of the Aleut kayak (*Baidarka*) to the evolution of digital computing and telecommunications (*Darwin Among the Machines*) and nuclear bomb–propelled space exploration (*Project Orion*).

During the years I spent kayaking along the coast of British Columbia and Southeast Alaska, I observed that the local raven populations spoke in distinct dialects. The divisions between these dialects appeared to correspond to the traditional geographic divisions between the indigenous human language groups. Ravens from Kwakiutl, Tsimshian, Haida, and Tlingit territory sounded different from one another, especially in their characteristic "tok" and "tlik."

I believe that this correspondence between human language and raven language represents coevolution rather than coincidence, though this would be difficult to prove.

Alison Gopnik

ALISON GOPNIK is a professor of cognitive science in the Psychology Department of the University of California at Berkeley. She is the author of, among other books, *The Scientist in the Crib: What Early Learning Tells Us About the Mind.*

I believe, but cannot prove, that babies and young children are actually *more* conscious, more vividly aware of their external world and internal life, than adults are. I believe this because there is strong evidence for a functional trade-off with development. Young children are much better than adults at learning new things and flexibly changing what they think about the world. On the other hand, they are much worse at using their knowledge to act in a swift, efficient, and automatic way. They can learn three languages at once, but they can't tie their shoelaces.

This trade-off makes sense from an evolutionary perspective. Our species relies more on learning than any other and has a longer childhood than any other. Human childhood is a protected period in which we are free to learn without being forced to act. There is even some neurological evidence for this. Young children have substantially more neural connections than adults—more potential to put different kinds of information together. With experience, some connections are strengthened and many others disappear entirely. As the neuroscientists say, we gain conductive efficiency but lose plasticity.

What does this have to do with consciousness? Consider the experiences we adults associate with these two kinds of functions. When we know how to do something really well and efficiently, we typically lose, or at least reduce, our conscious awareness of that action. We literally don't see the familiar houses and streets on the well-worn route home, although of course in some functional sense we must be visually taking them in. In contrast, when we are faced with the unfamiliar, when we fall in love with someone new, when we travel to a new place, our consciousness of what is around us and inside us suddenly becomes far more vivid and intense. In fact, we are willing to expend lots of money and emotional energy on those few intensely alive days in Paris or New York, which we will remember long after months of everyday life have vanished.

Similarly, when we as adults need to learn something new—say, when we learn to skydive, or when we work out a new scientific idea, or even when we are dealing with a new computer—we become vividly, even painfully, conscious of what we're doing; we need, as we say, to pay attention. As we become expert, we need to pay less and less attention, and we experience the movements and thoughts and keystrokes less and less. We sometimes say that adults are better at paying attention than children are, but really we mean just the opposite. Adults are better at *not* paying attention. They're better at screening out everything else and restricting their consciousness to a single focus. There is a certain amount of brain evidence for this too. Some brain areas, like the dorsolateral prefrontal cortex, consistently light up in adults when they are deeply engaged in learning something new. For more everyday tasks, these areas light up much less extensively. In children, the pattern is different—these areas light up even for mundane tasks.

The astute reader will note that this is just the opposite of what Dan Dennett believes but cannot prove. And this brings me to something else I believe but cannot prove. I believe that the problem of capital-C Consciousness will disappear in psychology just as the problem of Life disappeared in biology. Instead we'll develop much more complex, fine-grained, and theoretically driven accounts of the connections between particular types of phenomenological experience and particular functional and neurological phenomena. The vividness and intensity of our attentive awareness, for example, may be completely divorced from our experience of a constant first-person I. Babies may be more conscious in one way and less in the other. The consciousness of pain may be entirely different from the consciousness of red, which may be entirely different from the babbling stream of Joyce and Woolf.

Certainly, however, the vivid, even ecstatic awareness of the world that accompanies discovery is at least one kind of consciousness; indeed, it is the kind of consciousness that makes us grateful to be human. I think that for babies every wobbly step is skydiving, every game of hide-and-seek is Einstein in 1905, and every day is first love in Paris.

Paul Bloom

PAUL BLOOM is a professor of psychology at Yale University and the author of *Descartes' Baby: How the Science of Child Development Explains What Makes Us Human.*

The McGill psychologist John Macnamara once proposed that children come to learn about right and wrong, good and evil, in much the same way they learn about geometry and mathematics. Moral development is not merely cultural learning, and it does not reduce to the maturation of innate principles that have evolved through natural selection. It is not like the development of language or sexual preference or taste in food. Instead, moral development involves the construction of an intricate formal system that makes contact with the external world in a significant way.

This cannot be entirely right. We know that gut feelings, such as reactions of empathy or disgust, have a major influence on how children and adults reason about morality. And no serious theory of moral development can ignore the role of natural selection in shaping our moral intuitions. But what I like about Macnamara's proposal is that it allows for moral realism. It allows for the existence of moral truths that people discover, just as we discover truths of mathematics. We can reject the nihilist position (held by many researchers) that our moral intuitions are nothing more than accidents of biology or culture. And so I believe (though I cannot prove it) that the development of moral reasoning is the same sort of process as the development of mathematical reasoning.

William H. Calvin

WILLIAM H. CALVIN is a theoretical neurobiologist and an affiliate professor of psychiatry and behavioral sciences at the University of Washington in Seattle. He is the author of a dozen books, the latest of which is *A Brief History of the Mind: From Apes to Intellect and Beyond*.

Dan Dennett has it right when he puts the emphasis on acquiring language, not having language, as a precondition for our kind of consciousness. I have some (likely unprovable) beliefs about why the preschooler's acquisition of a structured language is so important for all the rest of her higher intellectual function. Besides syntax, intellect includes structured stuff such as multistage contingent planning, chains of logic, games with arbitrary rules, and our passion for discovering "how things hang together."

Many animals have some version of a critical period for tuning up sensory perception. Humans also seem to have one for structured language, judging from studies of deaf children with hearing parents who are not exposed to a rich sign language in the preschool years. In *Seeing Voices*, Oliver Sacks described an eleven-year-old boy who had been thought to be retarded but proved to be merely deaf. After a year of instruction in American Sign Language, Sacks interviewed him:

"Joseph saw, distinguished, categorized, used; he had no problems with perceptual categorization or generalization, but he could not, it seemed, go much beyond this, hold abstract ideas in mind, reflect, play, plan. He seemed completely literal—unable to juggle images or hypotheses or possibilities, unable to enter an imaginative or figurative realm. . . .He seemed, like an animal, or an infant, to be stuck in the present, to be confined to literal and immediate perception."

In the first year, an infant is busy creating categories for the speech sounds she hears. By the second year, the toddler is busy picking up new words, each composed of a series of phoneme building blocks. In the third year, she starts picking up on those typical combinations of words we call grammar or syntax. She soon graduates to speaking long structured sentences. In the fourth year, she infers a patterning to the sentences and starts demanding proper endings for her bedtime stories. It is pyramiding, using the building blocks at the immediately subjacent level. Four levels in four years!

These years see a lot of softwiring, with the pruning or the enhancement of prenatal connections between cortical neurons, depending partly on how useful a connection has been so far in life. Some connections help you assemble a novel combination of words, check them for nonsense via some sort of quality control, and then—*mirabile dictu*—speak a sentence you've never uttered before. Some of the connections must be in workspaces that could not only plan sentences but an agenda for the weekend, or negotiate a chain of logic, or assess a potential chess move—or even be tickled by structured music, with its multiple interwoven melodies.

Tuning up the workspace for structured language in the preschool years would likely carry over to those other structured aspects of intellect. That's why I like the emphasis on acquiring language as a precondition for consciousness: Tuning up to sentence structure might make a child better able to perform nonlanguage tasks that also need some structuring. Improve one, improve them all?

Is that what boosts our cleverness and intelligence? Is "our kind of consciousness" nothing but structured intellect with good quality control? Can't prove it, but it sure looks like a good bet.

Robert R. Provine

ROBERT R. PROVINE is a professor of psychology and neuroscience at the University of Maryland, Baltimore County. He is the author of *Laughter: A Scientific Investigation.*

Until it's proved otherwise, why not assume that consciousness does not play a significant role in human behavior? Although this idea may seem radical at first, it is actually the conservative position, the one that makes the fewest assumptions. The null position is an antidote to philosopher's disease—the inappropriate attribution of rational, conscious control over processes that may be irrational and unconscious. The argument is not that we lack consciousness but that we overestimate the conscious control of behavior.

I believe that statement to be true, but proving it is a challenge, because it's difficult to think about consciousness. We are misled by an inner voice that generates a reasonable but often fallacious narrative and explanation of our actions. That the beam of conscious awareness illuminating our actions is on only part of the time further complicates the task. Since we are not conscious of our state of unconsciousness, we vastly overestimate the amount of time that we are aware of our actions, whatever their cause.

My thinking about unconscious control was shaped by my field studies of the primitive play-vocalization of laughter. When I asked people to explain why they laughed in a particular situation, they would concoct some reasonable fiction about the cause of their behavior ("She did something funny," "It was something she said," "I wanted to put her at ease"). Observations of social context showed that such explanations were usually wrong. In clinical settings, such post-hoc misattributions would be termed "confabulations"—honest but flawed attempts to explain one's actions.

Subjects also incorrectly presumed that laughing is a choice and under conscious control—a reason for their confident, if bogus, explanations of their behavior. But laughing is not a matter of uttering "Ha-ha," as we would choose a word in speech. When challenged to laugh on command, most subjects could not do so. In certain, usually playful social contexts, laughter simply happens. However, this lack of voluntary control does not preclude an orderly, predictable pattern of behavior. Laughter appears at those places where punctuation would appear in the transcript of a conversation; it seldom interrupts the phrase structure of speech. We may say, "I have to go now—ha-ha" but rarely "I have to—ha-ha—go now." This punctuation effect is highly reliable and requires the coordination of laughing with the linguistic structure of speech, yet it is performed without the conscious awareness of the speaker. Other airway maneuvers, such as breathing and coughing, also punctuate speech and are performed without speaker awareness.

The discovery of structured but unconsciously controlled laughter produced by people who could not accurately explain their actions led me to consider generalizing this situation to other kinds of behavior. Do we go through life listening to an inner voice that provides similar confabulations about the causes of our action? Are essential details of the neurological process that governs human behavior inaccessible to introspection? Can the question of animal consciousness be stood on its head and treated in a more parsimonious manner? Instead of wondering whether other animals are conscious, or have a different, or lesser, consciousness than ours, should we be wondering whether our behavior is under no more conscious control than theirs? The complex social order of bees, ants, and termites documents what can be achieved with little if any conscious control, as we think of it. Is machine consciousness possible or even desirable? Is intelligent behavior a sign of conscious control? What kinds of tasks require consciousness? Answering these questions requires an often counterintuitive approach to the role, evolution, and development of consciousness.

Stanislas Dehaene

STANISLAS DEHAENE, the director of the Cognitive Neuroimaging Unit, Service Hospitalier Frédéric Joliot, Orsay, studies the cognitive neuropsychology of language and number processing in the human brain. He is the author of *The Number Sense: How the Mind Creates Mathematics*.

I believe (but cannot prove) that we vastly underestimate the differences that set the human brain apart from the brains of other primates.

Certainly, no one can deny that there are important similarities in the overall layout of the human brain and, say, that of the macaque. Our primary sensory and motor cortices are organized in similar ways; even in higher brain areas, homologies can be found. Using brain-imaging methods, my lab has observed plausible counterparts in the human parietal lobe—areas involved in eye movement, hand gestures, and number processing—to several areas of the macaque brain.

Yet I fear that those early successes in drawing human-monkey homologies have tended to mask notable differences. If we compare the primary visual regions of macaques and humans, there is already a two-fold difference in surface area; this ratio becomes a twenty- to fifty-fold increase in higher areas of the parietal and frontal lobes. Many of us suspect that in regions such as the prefrontal and inferior parietal cortices, the changes are so dramatic that they amount to additional brain areas. At a more microscopic level, there is a type of neuron reported to be found in the anterior cingulate region of humans and great apes but not in other primates; these so-called spindle cells send connections throughout the cortex, contributing to the much greater long-distance connectivity in the human brain. Such surface and connectivity differences, although they are in many cases purely quantitative, have brought about a qualitative revolution in brain function.

Jean-Pierre Changeux of the Institut Pasteur and I have proposed that the greater connectivity of the human brain enables a unique and flexible communication between distant brain areas. Human beings may have roughly the same specialized cerebral processors as our primate ancestors; however, what may be unique about the human brain is its ability to access the information inside each processor and make it available to almost any other processor through long-distance connections. I believe that we humans have a much more developed conscious workspace—a set of brain areas that can fluidly exchange signals, allowing us to internally manipulate information and perform unique mental syntheses. Using the workspace's long-distance connections, we can mobilize, in a top-down manner, essentially any brain area and bring it into consciousness.

Once the internal connectivity of a system exceeds a certain threshold, it begins to be dominated by self-sustained states of activity. I believe that the human workspace system has passed this threshold and gained a considerable autonomy: That is, the human brain is much less at the mercy of signals from the outside world than the brains of other primates are. Its activity never ceases to reverberate from area to area, generating a highly structured spontaneous flow of thoughts, which we project on the outside world.

Of course, spontaneous brain activity is present in all species, but if I am correct we will discover that it is both more evident and more structured in the human brain—at least in higher cortical areas, where "workspace" neurons with long-distance axons are denser. Furthermore, if human brain activity can be detached from outside stimulation, we will need to find new paradigms to study it, since bombarding the human brain with stimuli, as we do in most brain-imaging experiments, will not suffice. There is already some evidence for this: By comparing fMRI activations in humans and macaques evoked by the same visual stimuli, Guy Orban and his colleagues at the Catholic University of Leuven have found that prefrontal cortex activity is five times larger in macaques, noting that "there may be more volitional control over visual processing in humans than in monkeys."

The human species is also unique in its ability to expand its functionality by inventing new cultural tools. Writing, arithmetic, science—all are recent inventions. Our brains did not have enough time to evolve for them, but I speculate that they were made possible because we can mobilize our old areas in novel ways. When we learn to read, we recycle a specific region of our visual system known as the visual word-form area, enabling us to recognize strings of letters and connect them to language areas. Likewise, when we learn Arabic numerals we build a circuit to quickly convert those shapes into quantities—a fast connection from bilateral visual areas to the parietal quantity area. Even an invention as elementary as finger-counting changes our cognitive abilities dramatically. Amazonian people who have not invented counting are unable to make exact calculations as simple as, say, 6–2.

This "cultural recycling" implies that the functional architecture of the human brain results from a complex mixture of biological and cultural constraints. Education is likely to greatly increase the gap between the human brain and that of our primate cousins. Virtually all human-brain-imaging experiments today are performed on highly literate volunteers—and therefore, presumably, highly transformed brains. To better understand the differences between the human and the monkey brain, we need to invent new methods—both to decipher the organization of the infant brain and to study how it changes with education.

Stephen Kosslyn

STEPHEN KOSSLYN is a professor of psychology at Harvard University and an associate psychologist in the Department of Neurology at the Massachusetts General Hospital. He is the author (with Olivier Koenig) of *Wet Mind: The New Cognitive Neuroscience*.

These days, it seems obvious that the mind arises from the brain (not the heart, liver, or some other organ). In fact, I have gone as far as to claim that "the mind is what the brain does." But this notion does not preclude an unconventional idea: Your mind may arise not simply from your own brain but in part from the brains of other people.

Let me explain. This idea rests on three key observations.

The first is that our brains are limited, so we use crutches to supplement and extend our abilities. For example, try multiplying 756 by 312 in your head. Difficult, right? You would be happier with a pencil and a piece of paper—or, better yet, an electronic calculator. These devices serve as prosthetic systems, making up for cognitive deficiencies, just as a wooden leg would make up for a physical deficiency.

The second observation is that the major prosthetic system we use is other people. We set up what I call social prosthetic systems, or SPSs, in which we rely on others to extend our reasoning abilities and help us regulate and constructively employ our emotions. A good marriage may arise in part because two people can serve as effective SPSs for each other.

The third observation is that a key element of serving as an SPS is learning how best to help someone. Those who function as your SPSs adapt to your particular needs, desires, and predilections. And the act of learning changes the brain. By becoming your SPS, a person lends you part of his or her brain!

In short, parts of other people's brains come to serve as extensions of your own. And if the mind is what the brain does, then your mind arises from the activity of not only your own brain but also those of your social prosthetic systems.

There are many implications of these ideas, ranging from reasons why we behave in certain ways toward others to foundations of ethics and even of religion. In fact, one might argue that when your body dies, part of your mind may survive. But before getting into such dark and dusty corners, it would be nice to have firm footing—to collect evidence that these speculations are worth taking seriously.

Alex Pentland

ALEX (SANDY) PENTLAND is a pioneer in wearable computers, health systems, smart environments, and technology for developing countries. He is the Toshiba Professor of Media Arts and Sciences at MIT and currently directs the Human Dynamics research group at the MIT Media Lab.

What would it be like to be part of a distributed intelligence but still with an individual consciousness? Well, for starters you might expect the collective mind to take over from time to time, directly guiding the individual minds. Angry mobs and frightened crowds seem to qualify as examples of a collective mind in action, with nonlinguistic channels of communication usurping the individual capacity for rational behavior.

But as powerful as this sort of group compulsion can be, it is usually regarded simply as a failure of individual rationality, as a primitive behavioral safety net for the tribe in times of great stress. Surely this tribal mind doesn't operate in normal, day-today behavior—or does it? If human behavior were in substantial part due to a collective tribal mind, you would expect that nonlinguistic social signaling—the type that drives mob behavior—would be predictive of even the most rational and important human interactions. Analogous to the wiggle dance of the honeybee, there would be nonlinguistic signals that accurately predicted important behavioral outcomes.

And that is exactly what I find. Together with my research group, I have built a computer system that measures a set of nonlinguistic social signals, such as engagement, mirroring, activity, and stress, by analyzing "tone of voice" over one-minute periods. Although people are largely unconscious of this type of behavior, other researchers (Jaffe, Chartrand and Bargh, France, Kagen) have shown that similar measurements are predictive of infant language development and of empathy, depression, and even personality development in children. We have found that we can use these measurements of social signaling to predict a wide range of important behavioral outcomes with high accuracy.

Examples of objective and instrumental behaviors whose outcome we can accurately predict include salary negotiations, dating decisions, and roles in the social network. Examples of subjective predictions include hiring preferences and indications of empathy or interest. Accurate predictions can be made, even for lengthy interactions, by observing only the initial few minutes of the interaction, even though the linguistic content of these "thin slices" of the behavior seems to have little predictive power.

I find all of this astounding. We are examining some of the most important interactions a human being can have: finding a mate, getting a job, negotiating a salary, finding one's place in one's social network. These are activities for which we prepare intellectually and strategically, sometimes for decades, and yet the largely unconscious social signaling that occurs at the start of the interaction appears to be more predictive of its outcome than either the contextual facts (Is he attractive? Is she experienced?) or the linguistic structure (strategy chosen, arguments employed, and so forth).

So what is going on here? One might speculate that the social signaling we are measuring evolved as a method of establishing tribal hierarchy and cohesion, analogous to the psychologist Robin Dunbar's view that language evolved as grooming behavior. In this view, the tribal mind would function as unconscious collective discussion about relationships and resources, risks and rewards, and would interact with the conscious individual minds by filtering ideas according to their value relative to the tribe. Our measurements tap into the discussion and predict outcome by use of social regularities. For instance, in a salary negotiation it is important for the lower-status individual to establish that he or she is a "team player" by being empathetic, while in a potential dating situation the key variable is the female's level of interest. In our data there seem to be patterns of signaling that reliably lead to the desired states.

One question to ask about this social signaling is whether or not it is an independent channel of communication—that is, is it causal or do the signals arise from the linguistic structure? We don't have the full answer to that yet, but we do know that similar measurements predict infant language and personality development, and that adults can change their signaling by adopting different roles or identities within a conversation. Moreover, in our studies the linguistic and factual content seem uncorrelated with the pattern or intensity of the social signaling. So even if social signaling turns out to be only an adjunct to normal linguistic structure, it is a very interesting addition—a little like having speech annotated with speaker intent!

So here is what I suspect but cannot prove: A very large part of our behavior is determined by mainly unconscious social signaling, which sets the context, risk, and reward structure within which traditional cognitive processes proceed. This conjecture resonates with Steven Pinker's view of brain complexity and Stephen Kosslyn's thoughts about "social prosthetic systems." It also provides a concrete mechanism for the well-known processes of group polarization, groupthink, and the sometimes irrational behaviors of large groups. In short, it may be useful to start thinking of humans as having a collective, tribal mind in addition to personal ones.

Irene Pepperberg

IRENE PEPPERBERG is an adjunct professor of psychology at Brandeis University and a Bunting Fellow at the Radcliffe Institute for Advanced Study. The main focus of her work is to determine the cognitive and communicative abilities of grey parrots and compare their abilities with those of great apes, marine mammals, and young children. She is the author of *The Alex Studies*.

I believe, but can't prove, that human language evolved from a combination of gesture and innate vocalizations, via the concomitant evolution of mirror neurons, and that birds will provide the best model for language evolution.

Work on mirror neurons—that is, neurons that fire both when one performs a particular action and when one observes another performing it—over the past decade has provided intriguing evidence (although no solid proof) for the gestural origins of speech. What can be called the mirror-neuron hypothesis suggests that only a small reorganization of the nonhuman primate brain was needed to create the wiring that underlies speech acquisition/learning. What is missing from the hypothesis is a model of the development of language from speech. It is here that I believe that a model based on avian vocalizations is most valuable.

First, some background. Passerine birds can be divided into two groups: the oscines, who learn their songs, and the sub-oscines, who have a limited number of what seem to be innately specified songs. The former have well-defined neural architectures and mechanisms for song acquisition; the latter lack brain structures for song acquisition, although they obviously have brain and vocal-tract structures for producing song. The sub-oscines, in parallel with nonhuman primates, often use various behaviors or gestures (posture, numbers of song repetitions, feather erectness, types of flights, and so on) to provide additional information about the meaning of their utterances. W. John Smith can predict, for example, a flycatcher's actions by the combination of posture, flight, and singing pattern he observes. The songbirds, like human children learning language, will not learn their vocalizations if deafened, and need to hear, babble, and practice songs before attaining adult competence. Very recent work by G. J. Rose and his colleagues demonstrates that even the syntax of their song is learned, through early exposure to paired phrases, which are then combined to create the adult vocalizations. Such data, demonstrating how sparrows integrate information about temporally related events and how they use that information to develop sequential vocal behavior, amount to a viable model for human syntax acquisition.

Now, no one knows whether any birds have mirror neurons, or how their mirror neurons would function if they did; some neural data on a bird's responses to its own song (as played back to it, not what it hears while it is singing) provide intriguing hints. I predict (a) the existence of such neurons in oscines, and (b) that such neurons will have a robust role in oscine song development, but (c) that only more primitively functioning mirror neurons (akin to the differences between monkey and human mirror neurons) will be found in sub-oscines.

What about the so-called missing link between learned and unlearned vocal behavior? No one has found such a missing link in the primate line, but Donald Kroodsma has recently discovered a flycatcher (a supposedly sub-oscine bird) that apparently learns its song. The song is simple but has variations among groups of birds that constitute dialects. No one yet knows if these birds have brain mechanisms for song learning or what these mechanisms might be. But I predict that Kroodsma's flycatchers will be found to have mirror neurons that function in an intermediate manner between those of the oscines and sub-oscines and will provide a model for the missing link between nonhuman primate and human communication.

Howard Gardner

HOWARD GARDNER is the John H. and Elisabeth A. Hobbs Professor in Cognition and Education at the Harvard Graduate School of Education. He is also adjunct professor of psychology at Harvard University and adjunct professor of neurology at the Boston University School of Medicine. Among his most recent books are *The Disciplined Mind*, *Intelligence Reframed*, and *Changing Minds*.

I believe that human talents are based on distinct patterns of brain connectivity. These patterns can be observed as the individual encounters and ultimately masters an organized activity or domain in his or her culture.

Consider three competing accounts:

1. Talent is a question of practice. We could all become Mozarts or Einsteins if we persevered.
2. Talents are fungible. A person who is good in one thing could be good in everything.
3. The basis of talents is genetic.While true, this account misleadingly implies that people with a "musical gene" will necessarily evince their musicianship, just as they evince their eye color or, less happily, Huntington's disease.

My account: The most apt analogy is language learning. Nearly all of us can easily master natural languages in the first years of life; we might say that nearly all of us are talented speakers. An analogous process occurs with respect to various talents, with two differences:

a. There is greater genetic variance in the potential to evince talent in areas like music, chess, golf, mathematics, leadership, written (as opposed to oral) language, and so on.

b. Compared to language, the set of relevant activities is more variable within and across cultures. Consider the set of games. A person who masters chess easily in one culture would not necessarily master poker or "go" in another.

As we attempt to master an activity, neural connections of varying degrees of utility or disutility form. Certain of us have nervous systems predisposed to develop quickly along the lines needed to master specific activities (chess) or classes of activities (mathematics) that happen to be available in one or more cultures. Accordingly, assuming such exposure, we will appear talented and become experts quickly. The rest of us can still achieve some expertise, but it will take longer, require more effective teaching, and draw on intellectual faculties and brain networks that the talented person does not have to use.

This hypothesis is currently being tested by Ellen Winner and Gottfried Schlaug. These investigators are imaging the brains of young students before they begin music lessons and for several years thereafter. They also are imaging control groups and administering control (non-musical) tasks. After several years of music lessons, judges will determine which students have musical "talent." The researchers will document the brains of musically talented children before training and how these brains develop.

If account No.1 is true, hours of practice will explain all. If No.2 is true, those best at music should excel at all activities. If No.3 is true, individual brain differences should be observable from the start. If my account is true, the most talented students will be distinguished not by differences observable prior to training but rather by the ways in which their neural connections alter during the first years of training.

David Gelernter

DAVID GELERNTER is a professor of computer science at Yale and chief scientist at Mirror Worlds Technologies in New Haven. His research centers on information management, parallel programming, and artificial intelligence. He is the author of *Mirror Worlds*, *The Muse in the Machine*, and *Drawing a Life: Surviving the Unabomber*.

I believe (I *know*—but can't prove!) that scientists will soon understand the physiological basis of the cognitive spectrum, from the bright violet of tightly focused analytic thought all the way down to the long, slow red of low-focus sleep thought—also known as dreaming. Once they understand the spectrum, they'll know how to treat insomnia, and they will understand analogy-discovery (and therefore creativity) and the role of emotion in thought—and that thought takes place not only when you solve a math problem but also when you look out the window and let your mind wander. Computer scientists will finally understand the missing mystery ingredient that made all their efforts to simulate human thought such naive, static failures and turned this once thriving research field into a ghost town. (Their failures were static insofar as people think in different ways at different times: Your energetic, wide-awake mind works very differently from your tired, soon-to-be-sleeping mind, but artificial intelligence programs always "thought" in the same way, all the time.)

And scientists will understand why we can't force ourselves to fall asleep or be creative—and how those two facts are related. They'll understand why so many people report being most creative while driving, shaving, or doing some other activity that keeps the mind's foreground occupied and lets it approach open problems in a low-focus way. In short, they'll understand the mind as an integrated, *dynamic* process that changes over a day and a lifetime but is characterized always by one continuous spectrum.

Here's what we know about the cognitive spectrum: Every human being traces out some version of the spectrum every day. You're most capable of analysis when you are most awake. As you grow less wide-awake, your thinking grows more concrete. As you start to fall asleep, you begin to free-associate. (Cognitive psychologists have known for years that you begin to dream before you fall asleep.) We know also that to grow up intellectually means to trace out the cognitive spectrum in reverse: Infants and children think concretely; as they grow up, they're increasingly capable of analysis. (Not incidentally, newborns spend nearly all their time asleep.)

Here's what we suspect about the cognitive spectrum: As you move down the spectrum, as your thinking grows less analytical and more concrete and finally bottoms on the wholly nonlogical, highly concrete type of thought we call "dreaming," emotions function increasingly as the "glue" of thought. I can't prove (but I believe) that "emotion coding" explains the problem of analogy. Scientists and philosophers have knocked their heads against this particular brick wall for years: How can people say, "A brick wall and a hard problem seem wholly different, yet I can draw an analogy between them"? If we knew the answer to that, we'd understand the essence of creativity. The answer is:

We are able to draw an analogy between two seemingly unlike things because the two *are associated in our minds with the same emotion.* And that emotion acts as a connecting bridge between them. Each memory comes with a characteristic emotion; similar emotions allow us to connect two otherwise unlike memories. An emotion isn't the crude, simple thing we make it out to be in speaking or writing—"happy," "sad," and so on; an emotion can be the delicate, complex, nuanced, inexpressible feeling you get on the first warm day in spring.

And here's what we don't know: What's the physiological mechanism of the cognitive spectrum? What's the genetic basis? Within a generation, we'll have the answers.

Marc D. Hauser

MARC D. HAUSER is a professor of psychology in Harvard University's Psychology Department and codirector of the Mind, Brain, and Behavior Program. He is the author of *The Evolution of Communication*, *The Design of Animal Communication*, and *Wild Minds*.

What makes humans uniquely smart?

Here's my best guess: We alone evolved a simple computational trick with far-reaching implications for every aspect of our life, from language and mathematics to art, music, and morality. The trick: the ability to take as input any set of discrete entities and recombine them into an infinite variety of meaningful expressions.

Thus we take meaningless phonemes and combine them into words, words into phrases, and phrases into Shakespeare. We take meaningless strokes of paint and combine them into shapes, shapes into flowers, and flowers into Monet's water lilies. And we take meaningless actions and combine them into action sequences, sequences into events, and events into homicide and heroic rescues.

I'll go one step further: I bet that when we discover (intelligent) life on other planets, we'll find that although the materials may be different for running the computation, they will create open-ended systems of expression by means of the same trick, thereby giving birth to the process of universal computation.

Gary Marcus

GARY MARCUS is associate professor of psychology and neural science at New York University and director of its Infant Language Center. He is the author of *The Algebraic Mind* and *The Birth of the Mind*.

If computers are made up of hardware and software, transistors and resistors, what are the neural machines that we know as minds made of? Although it doesn't take a neuroscientist to realize that minds aren't literally made of transistors and resistors, I firmly believe that human machines share one of the most basic elements of computation: the ability to represent information in terms of an abstract, algebralike code.

Virtually all of the world's computer software consists of thousands (or even millions) of instructions that say things like "If X is greater than Y, do Z" or "Calculate the value of *Q* by adding *A*, *B*, and *C*," thereby yielding recipes that work not just for specific cases but for enormous ranges of possible data sets—anything that can be plugged into variables like X, Y, or Z.

My contention is that human cognitive systems rely on much the same type of abstraction. For example, the famous linguistic dictum that a sentence can be formed by combining any noun phrase with any verb phrase generates not just Noam Chomsky's celebrated *Colorless green ideas sleep furiously* but a potentially infinite number of sentences. In its open-endedness, language is a paradigmatic example of mental algebra and the flexibility of human thought.

Remarkably, even human infants seem capable of this kind of abstraction. In my lab, we played seven-month-old babies a series of made-up sentences like *la ta la, ga na ga, je li je* (generated from what we called an ABA grammar) or *la ta ta, ga na na, je li li* (an ABB grammar). In just two minutes, the babies were able to pick up these simple but abstract grammars, strongly suggesting that infants are born with a vital capacity for algebraic representation.

This strong suggestion awaits final confirmation—behavioral studies can only hint at the underlying neural mechanism. In the final analysis, we will need not-yet-invented neuroscientific techniques that take us to the level of understanding interactions between individual neurons. But every bit of evidence we can collect now—from babies, from toddlers, from adults, from psychology, and from linguistics—point to the idea that the algebralike abstractions that make computers tick play an equally important role in human neurocognitive systems.

Brian Goodwin

BRIAN GOODWIN is a professor of biology at Schumacher College, in Devon, U.K., and the author of *How the Leopard Changed Its Spots*.

I believe that nature and culture can be understood as one unified process, not two distinct domains separated by some property of human beings, such as written or spoken language, consciousness, or ethics.

Although there is no proof of this, and no consensus in the scientific community or in the humanities, revelations of the past few years provide a foundation for both empirical and conceptual work that I believe will lead to a coherent, unified perspective on the process in which we and nature are engaged. This is not a takeover of the humanities by science but a genuine fusion of the two, based on articulations of fundamental concepts, such as meaning and wholeness in natural and cultural processes, and with implications for scientific studies, their applications in technology, and their expression in the arts.

For me, this vision arrives primarily through developments in biology, which occupies the middle ground between culture and the physical world. The key conceptual changes have arisen from complexity theory, with its detailed studies of interactions between the components of organisms and also between organisms in ecosystems. When the genome projects made it clear that we cannot make sense of the information in DNA, attention shifted to understanding how organisms use this information to make themselves with forms that allow them to survive and reproduce in particular habitats. The focus turned from the hereditary material to its organized context, the living cell; organisms as agencies with a distinctive kind of organization returned to the biological foreground.

Examinations of the self-referential networks that regulate gene activities in organisms, carrying out the diverse functions and constructions within cells through protein-protein interactions, and of the sequences of an organism's metabolic transformations reveal that they all have distinctive properties of self-similar fractal structure, governed by power-law relationships.

These properties are similar to the structure of languages, which are also self-referential networks described by power laws (as discovered years ago by the Harvard linguist G. K. Zipf). A conclusion is that organisms use proto-languages to make sense both of their inherited history (written in DNA and its molecular modifications) and of their external context (the environment), in the process of making themselves as functional agents. Organisms thus become participants in cultures with histories that have meaning, expressed in the forms (morphologies and behaviors) distinctive to their species. This is embodied or tacit meaning, which cognitive scientists now recognize as also primary in human culture.

Understanding species as cultures that have experienced 3.7 billion years of adaptive evolution on Earth makes it clear that they are repositories of meaningful knowledge and experience about effective living which we urgently need to learn about in human culture. Here is a source of deep wisdom about living in participation with others in ways that are energy- and resource-efficient, that recycle everything, that produce forms both functional and beautiful, and that are continuously innovative and creative. We can now proceed with a holistic science unified with the arts and humanities, which has as its foundation the principles of a naturalistic ethic based on an extended science that includes qualities as well as quantities.

There is plenty of work to do in articulating this unified perspective—from empirical studies of how organisms achieve their states of coherence and adaptability to the application of these principles in the organic design of all human artifacts, from energy-generating devices and communication systems to cars and factories. The goal is to make human culture as integrated with natural processes as the rest of the living realm, so that we enhance the quality of the planet instead of degrading it. This will require a rethinking of evolution in terms of the intrinsic agency with meaning embodied in the life cycles of different species, understood as natural cultures.

Integrating biology and culture with physical principles will be something of a challenge, but there are already indications of how it can be achieved. For example, the self-similar fractal patterns that arise in physical systems during phase transitions, when new order is coming into being, have the same characteristics as the patterns observed in organismic and cultural networks involved in generating order and meaning. The unified vision of a creative and meaningful cosmic process may well replace the meaningless mechanical cosmos that has dominated Western scientific thought and cultural life for the past few hundred years.

Leo M. Chalupa

LEO M. CHALUPA is Distinguished Professor of Ophthalmology and Neurobiology and chair of the Section of Neurobiology, Physiology, and Behavior at the University of California, Davis. His principal research interests concern the development and plasticity of the mammalian visual system and factors regulating the establishment of functional, neurochemical, and structural properties of retinal ganglion cells.

Here are three of my unproved beliefs:

1. The human brain is the most complex entity in the known universe;
2. With this marvelous product of evolution we will eventually succeed in discovering all there is to discover about the physical world—provided of course that some catastrophic event doesn't terminate our species; and
3. Science provides the best means of attaining this ultimate goal.

When the scientific endeavor is considered in relation to the obvious limitations of the human brain, the knowledge we have gained in all fields is astonishing. Consider the well-documented variability in the functional properties of neurons. When recordings are made of responses from a single neuron—for example, in the visual cortex, to a flashing spot of light—one can't help but be amazed by the trial-to-trial variations. In one instance, this simple stimulus might elicit a high-frequency burst of discharges, while in the next trial there might be just the hint of a response. The same phenomenon obtains in EEG recordings: Brain waves change in frequency and amplitude in seemingly random fashion, even when the subject is lying prone with no variation in behavior or environment. Such variability is also evident in brain imaging; the pretty pictures of brain states seen in publications are averages of many trials that have been massaged by computer.

So how does the brain do it? How can it function as effectively as it does, given the noise inherent in the system? I don't have a good answer and neither does anyone else, in spite of the papers that have been published on this problem. But in line with the second of the three beliefs above, I am certain that someday this question will receive a definitive answer.

Margaret Wertheim

MARGARET WERTHEIM, an internationally noted science writer and commentator, has written extensively about science and society for magazines, television, and radio. She is the author of *Pythagoras' Trousers* and *The Pearly Gates of Cyberspace* and is currently working on a book about the role of imagination in theoretical physics.

We all believe in something, and science itself is premised on a whole set of beliefs. Above all, science is founded on the belief that things are comprehensible and that by the ingenuity of our minds and the probing of ever more subtle instruments we will ultimately come to know It All.

But is the All inherently knowable? I believe, though I cannot prove it, that there will always be things we do not know—large things, small things, interesting things, and important things.

If theoretical physics is any guide, we might suppose that science is a march toward a finite goal. For the past few decades, theoretical physicists have been searching for the so-called Theory of Everything—what Nobel laureate Steven Weinberg has also called a "final theory." This ultimate set of equations would tie together all the fundamental forces that physicists recognize today: gravity, electromagnetism, and the nuclear forces inside the cores of atoms. But such a theory, if we are lucky enough to extract it from the current mass of competing contenders, would not tell us anything about how proteins form or how DNA came into being. Less still would it illuminate the machinations of a living cell or the workings of the human mind. A Theory of Everything would not even help us to understand how snowflakes form.

In an age when we have discovered the origin of the universe and observed the warping of space and time, it is shocking to hear that scientists do not understand something as seemingly paltry as the formation of ice crystals. But that is indeed the case. Caltech physicist Kenneth Libbrecht is a world expert on ice-crystal formation, a hobby/project he took on more than twenty years ago precisely because, as he puts it, "there are six billion people on this planet, and I thought at least one of us should understand how snow crystals form." After two decades of meticulous experimentation inside specially constructed pressurized chambers, Libbrecht believes he has made some headway in understanding how ice crystallizes at the edge of the quasi-liquid layer that surrounds all ice structures. He calls his theory "structure dependent attachment kinetics," but he is quick to point out that this is far from the ultimate answer. The transition from water to ice is a mysteriously complex process that has engaged minds as brilliant as Johannes Kepler and Michael Faraday. Libbrecht hopes he can add the small next step in our knowledge of this wondrous substance that is central to life itself.

Libbrecht is also one of hundreds of physicists working on the Laser Interferometer Gravitational-Wave Observatory (LIGO), which is designed to detect gravitational waves thought to emanate from black holes and other massive cosmological objects. Gravity waves are predicted by the general theory of relativity, hence physicists believe they must exist. Here belief has brought into being a half-billion-dollar machine. Any successful

Theory of Everything will have to account for gravity, which, along with the three other forces, should ultimately manifest itself in both wave and particle form. LIGO is designed to detect the wavelike face of this most mysterious force, if indeed that exists.

Some years ago the science writer John Horgan wrote a provocative book in which he suggested that science was coming to an end, all the major theoretical edifices now supposedly being in place. Horgan was right in one sense, for high-energy physics may be on the verge of achieving its final unification. But in so many other areas, science is just beginning to find its way. Only now, for example, are we acquiring the scientific tools and techniques to begin to investigate how our atmosphere works, how ecological systems function, how genes create proteins, how cells evolve, how brains work. The very success of "fundamental science" has opened doors closed to earlier generations—yet increasingly it seems there is more than ever that we do not know. At a time when the physics journals are full of theories about how to create entire universes in the laboratory, it is easy to imagine that science has grasped the whole of reality. In truth, our ignorance is vast—and I believe it will always be so.

Just before the flowering of the scientific revolution, the great early champion of mathematical science Cardinal Nicholas of Cusa advocated the advancement of what he termed "learned ignorance." Not omniscience but an ever more subtle and insightful *un-knowing* was the goal Cusa envisioned for the modern scientific mind. In the humble snowflakes Ken Libbrecht studies we have the metaphor for this inspiring view: Though they melt on your tongue, each tiny crystal of ice encapsulates a universe whose basic rules we have barely begun to discern.

Gino Segrè

GINO SEGRÈ, a noted theoretical physicist from a distinguished family of physicists, is a professor in the Department of Physics and Astronomy at the University of Pennsylvania. He is also the author of *A Matter of Degrees: What Temperature Reveals About the Past and Future of Our Species, Planet, and Universe.*

The Big Bang, that primeval explosion more than 13 billion years ago, provides the accepted description of our universe's beginning. We can trace with exquisite precision what happened during the expansion and cooling that followed that cataclysm, but the presence of neutrinos in the earliest phase continues to elude direct experimental confirmation.

Neutrinos, once they were in thermal equilibrium, were supposedly freed from their bonds to other particles about two seconds after the Bang. Since then, they should have been roaming undisturbed through intergalactic space, some 200 of them in every cubic centimeter of our universe, altogether a billion of them for every single atom. Their presence is noted indirectly in the universe's expansion; however, though they are presumably by far the most numerous type of material particle in existence, not a single one of those primordial neutrinos has ever been detected. It is not for want of trying, but the necessary experiments are almost unimaginably difficult. And yet those neutrinos must be there. If they are not, our picture of the early universe will have to be totally reconfigured.

Wolfgang Pauli's original 1930 proposal of the neutrino's existence was so daring that he didn't publish it. Enrico Fermi's brilliant 1934 theory of how neutrinos are produced in nuclear events was rejected for publication by *Nature* as too speculative. In the 1950s, neutrinos were detected in nuclear reactors and soon afterward in particle accelerators. Starting in the 1960s, an experimental tour de force revealed their existence in the solar core. Finally, in 1987, a ten-second burst of neutrinos was observed radiating outward from a supernova that occurred almost 200,000 years ago. When they reached Earth and were observed, one prominent physicist quipped that extrasolar neutrino astronomy "has gone in ten seconds from science fiction to science fact." These are some of the milestones of twentieth-century neutrino physics.

In the twenty-first century, we eagerly await another milestone—the observation of neutrinos produced in the first seconds after the Big Bang. We have been able to infer their presence, but will we be able to actually detect these minute and elusive particles? They must be everywhere around us, even though we still cannot prove it.

Haim Harari

HAIM HARARI, a theoretical physicist, is a former president (1988 to 2001) of the Weizmann Institute of Science. He is currently the chair of its Davidson Institute of Science Education.

The electron has been with us for over a century, laying the foundations for the electronic revolution and all of information technology. It is believed to be a pointlike, elementary, and indivisible particle. Is it?

The neutrino, more than a million times lighter than the electron, was predicted in the 1920s and discovered in the 1950s. It plays a crucial role in the creation of the stars, the sun, and the heavy elements. It is elusive, invisible, and weakly interacting. It is also considered fundamental and indivisible. Is it?

Quarks do not exist as free objects, except at extremely tiny distances deep within the confines of the particles—protons and neutrons, the constituents of atomic nuclei—that are constructed from them. Since the 1960s, we have believed that quarks are indivisible and the most fundamental nuclear building blocks. Are they?

Nature has created two sets of additional, totally unexplained replicas of the electron, the neutrino, and the two most abundant quarks (the up quark and the down quark). Each set is identical to the other two in all properties, except that the particle masses are radically different. Since each set includes four fundamental particles, we end up with twelve different particles, which are allegedly indivisible, pointlike, and elementary. Are they?

The atom, the atomic nucleus, and the proton, each in its own time, were considered elementary and indivisible, only to be subdivided later into new fundamental building blocks. How can we be so arrogant as to exclude the possibility that this will happen again? Why would nature arbitrarily produce twelve different objects, with a very orderly pattern of electric charges and "color forces," with simple charge ratios between seemingly unrelated particles (such as the electron and the quark), and with a pattern of masses that appears to be taken from the results of a lottery? Doesn't this smell of further subparticle structure?

There is absolutely no experimental evidence for a further substructure within all of these particles. There is no completely satisfactory theory that might explain how such light and tiny particles can contain objects moving with enormous energies, a requirement of quantum mechanics. This is presumably why the accepted view—the "party line"—among particle physicists is that we have already reached the most fundamental level of the structure of matter.

For more than twenty years, the hope has been that the rich spectrum of so-called fundamental particles will be explained as various modes of string vibrations and excitations. The astonishingly tiny string or membrane, rather than the pointlike object, is allegedly at the bottom of the ladder describing the structure of matter. However, in spite of absolutely brilliant and ingenious mathematical work, not one experimental number has been explained by the string hypothesis.

Based on common sense and on an observation of the pattern of the known particles, without any experimental evidence and without any comprehensive theory, I have believed for many years, and I continue to believe, that the electron, the neutrino, and the quarks are divisible. They are presumably made of different combinations of the same small number (two?) of more fundamental subparticles. The latter may or may not have the string structure, and may or may not be themselves composites.

Will we live to see the components of the electron?

Donald I. Williamson

DONALD I. WILLIAMSON is a biologist at the Port Erin Marine Laboratory of the University of Liverpool (U.K.). He is the author of *The Origins of Larvae*.

I believe I can explain the Cambrian explosion.

The Cambrian explosion refers to the appearance, in a relatively short space of geological time beginning more than 500 million years ago, of a very wide assortment of animals. I believe that it came about through hybridization.

Many well-preserved Cambrian fossils occur in the Burgess Shale of the Canadian Rockies. These fossils include small, soft-bodied animals, several of which are planktonic but none of which are larvae. Some of them seem to have the front end of one animal and the rear end of another. Modern larvae present a comparable set-up: that is, larvae seem to be derived from animals belonging to groups different from those of their adult form. I have amassed a bookful of evidence that the basic larval forms did indeed originate as animals in other groups, and that such forms were transferred by hybridization. Animals with a larval stage are "sequential chimeras"—creatures whose initial body-form, the larva, is followed by a distantly related form, the adult. I believe that there were no true Cambrian larvae, only pseudolarvae, in which the initial form became the front end of the adult. Cambrian hybridizations produced "concurrent chimeras," each made up of two distantly related body-forms.

About 600 million years ago, shortly before the Cambrian period, animals with tissues (metazoans) made their first appearance. I agree with Darwin that there were several different forms of metazoan—Darwin suggested four or five—and I believe that they resulted from hybridization between various colonial protists. (Some protists, single-celled animals, often occur in colonies, consisting of many similar cells.) All Cambrian animals were marine, and like most marine animals today, they shed their eggs and sperm into the water, where fertilization took place. Eggs of one species frequently encountered sperm of another species, and mechanisms to prevent hybridization would have been only poorly developed. Early animals had small genomes, which left their cells with plenty of spare gene capacity. These factors led to many fruitful hybridizations, which resulted in concurrent chimeras. Not only did the original metazoans hybridize but the new animals resulting from these hybridizations also hybridized, and this produced the explosion in animal forms.

The hybridizations that produced the first larvae came much later, when there was little room left in cell nuclei for more genes, and this process is still going on. Echinoderms (the group that includes sea urchins and starfish) had no larvae in either the Cambrian or the Ordovician (the following period), and this may well apply to other major faunal groups. Acquiring parts (instead of larvae) by hybridization continued, I believe, throughout those two periods and probably later, but as genomes became larger and filled most of the available nuclear space, such later hybridizations led to smaller and smaller changes in adult form—or else to the acquisition of larvae. The gradual evolution of better mechanisms to prevent eggs from being fertilized by foreign sperm resulted in fewer fruitful hybridizations, but occasional hybridizations still take place.

Hybridogenesis (the generation of new organisms by hybridization) and symbiogenesis (the generation of new organisms by symbiosis) are both rapid processes involving fusions of lineages, whereas Darwinian "descent with modification" is gradual and within separate lines of descent. These forms of evolution proceed in parallel, and natural selection works on the results.

I cannot prove that Cambrian animals had poorly developed specificity and spare gene capacity, but it makes sense.

Ian Wilmut

IAN WILMUT is a group leader in the Department of Gene Function and Development at the Roslin Institute, near Edinburgh, and is the leader of the team that in 1996 produced Dolly the sheep, the first animal to be cloned from an adult cell. He is the author, with Keith Campbell and Colin Tudge, of *The Second Creation*.

I believe that it is possible to change adult cells from one phenotype to another.

The birth of Dolly, the first animal to be cloned from an adult (of any species), provided the insight behind this belief. Previously, biologists had thought that the mechanisms directing the formation of the various tissues making up an adult were so complex and so rigidly fixed that they could not be reversed. Dolly's birth demonstrated that the mechanisms active in a nucleus transferred from a mammary epithelial cell could be reversed in the recipient unfertilized egg.

We take for granted the process by which a single-celled embryo gives rise to all of the many different tissues of an adult. Because almost all adult cells have the exact same genetic information, the differences among them must have arisen from sequential differences in the function of the genes. We are beginning to learn something of the factors promoting these sequential changes, although very little is known of the hierarchy of their influence. I believe that a greater understanding of these mechanisms will allow us to cause cells from one kind of adult tissue to form another kind of tissue.

We have long been accustomed to the idea that cells are influenced by their external environment, and in the laboratory we use specific methods of tissue culture to control their function. We will learn how to increase the activity of these "intracellular" factors—perhaps by direct introduction of proteins, or by the use of small-molecule drugs to modulate the expression of key regulatory genes or stimulate their transient expression. We have much to learn about the optimal approach to take: Is it necessary to reverse the differentiation process at an early stage in a particular pathway, for example, or can we achieve "transdifferentiation" directly from one pathway to another? The answer may vary depending on the kind of tissue, and the medical implications will be profound. Cells of specific tissues will be available from patients, either for the study of genetic differences or for their own therapy.

All this is not to suggest that we cease research on embryonic stem cells, because such investigations are essential to the development of the new methods I envisage; conversely, understanding the mechanisms of reprogramming cells will create important new opportunities in the use of embryonic stem cells. As many options as possible should be available to the researcher and clinician.

It is my belief that research into tissue formation will ultimately prove the most valuable legacy of the Dolly experiment. The ramifications are far wider than the mere production of cloned offspring.

Daniel Goleman

DANIEL GOLEMAN is a psychologist who for many years reported on the brain and behavioral sciences for the *New York Times*. He is the author of *Emotional Intelligence*.

I believe but cannot prove that today's children are unintended victims of economic and technological progress.

To be sure, greater wealth and advanced technology offer all of us better lives, yet these unstoppable forces seem to have transformed childhood in disastrous ways. Even as the average IQ of American children has steadily increased over the last century, the past three decades have seen a major drop in their most basic social and emotional skills—the very abilities they need to become effective workers and leaders, parents and spouses, members of the community.

There are always individual exceptions, but the bell curve for social and emotional abilities seems to be sliding in the wrong direction. The most compelling data come from a random nationwide sample, conducted by Thomas Achenbach at the University of Vermont, of more than 3,000 representative American schoolchildren aged seven to sixteen, whose behavior was rated by their parents and teachers—adults who knew them well. The first sampling was taken in the early 1970s, another roughly fifteen years later, and a third in the late

1990s. The results show a startling decline in social and emotional health.

There is a precipitous drop between the first and second cohorts. American children in the mid-1980s were more withdrawn, sulky, unhappy, anxious and depressed, impulsive and unable to concentrate, delinquent and aggressive, than they were in the early 1970s. They did worse on forty-two indicators, better on none. In the late 1990s, scores crept back up, but not as high as they had been on the first round.

That's the data. What I believe but can't prove is that this decline is due in large part to economic and technological forces. The ratcheting upward of global competition means that over the last two decades or so, parents have had to work longer to maintain the same standard of living their own parents enjoyed. Virtually every American family nowadays has two working parents; fifty years ago, one working parent was the norm. It's not that today's parents love their children less but that they have less free time to spend with them.

Increasing mobility means that fewer children live in the same neighborhood as their extended families and so no longer have surrogate parenting from close relatives. Day care can be excellent, particularly for children of privileged families, but too often the less well-to-do children get too little caring attention in their day. Middle-class childhood has become overly organized, a tight schedule of dance or piano lessons and soccer games, with children shuttled from one adult-run activity to another, making for less free time in which they can play together on their own, in their own way. When it comes to learning social and emotional skills, the loss of free time with family, relatives, and other children translates into a loss of the very activities that traditionally fostered the natural transmission of those skills.

Then there's the technological factor. Today's children—in the developed, and increasingly in the developing, world—

spend more time than ever in human history alone, staring at a video monitor. These circumstances amount to a natural experiment in child rearing on an unprecedented scale. While such children may grow up to be more at ease with computers, they are undoubtedly failing to acquire those skills that will enable them to relate to other human beings.

The prefrontal-limbic neural circuitry crucial to the acquisition of social and emotional abilities is the last part of the human brain to become anatomically mature, a developmental task not completed until the mid-twenties. During that window of time, the life abilities of a child become set, as neurons come online and are interconnected for better or worse. It is the experiences of one's childhood that dictate how those connections are made. A smart strategy for helping today's kids accomplish the right kind of social and emotional skill building would be to bring such lessons into the classroom rather than leaving them to chance in today's hard-driving technology-ridden wonderworld.

Esther Dyson

ESTHER DYSON is editor of Release 1.0 at CNET Networks and responsible for its PC Forum annual conference. From 1998 to 2000, she was founding chairman of ICANN, the organization overseeing Domain Name System policy. She is the author of *Release 2.0: A Design for Living in the Digital Age* and is an active investor in information technology start-ups.

We're living longer and thinking shorter.

It's all about time.

Modern life has fundamentally and paradoxically changed our sense of time. Even as we live longer, we seem to think shorter. Is it because we cram more into each hour, or because the next person over seems to cram more into each hour? For a variety of reasons, everything is happening much faster, and more things are happening. Change is a constant.

It used to be that machines automated work, giving us more time to do other things, but now machines automate the production of attention-consuming information, which *takes* our time. For example, if one person sends the same e-mail message to ten people, then ten people (in theory) should give it their attention. And that's a low-end example.

The physical friction of everyday life—the time it took Isaac

Newton to travel by coach from London to Cambridge, the dead spots of walking to work (no iPod), the darkness that kept us from reading—has disappeared, making every minute not used productively into an opportunity lost.

And finally, we can measure more, over smaller chunks of time. From airline miles to calories (and carbs and fat grams), from friends on Friendster to steps on a pedometer, from real-time stock prices to millions of burgers consumed, we count things by the minute and the second. Unfortunately, this carries over into how we think and plan: Businesses focus on short-term results; politicians focus on elections; school systems focus on test results; most of us focus on the weather rather than on the climate. Everyone knows about the big problems, but their behavior focuses on the here and now.

I first noticed this phenomenon full-fledged in the United States right after 9/11, when it became impossible to schedule an appointment or get anyone to make a commitment. To me, it felt like Russia, where I had been spending time since 1989; there, people had avoided making long-term plans because there was little discernible relationship between effort and result. Suddenly, even in the United States, people were behaving like the Russians of those days. Companies suspended their investments; individuals suspended their plans for new jobs, marriages, new houses . . . all activity slowed; everything became "I'll consider" or "I'll try," rather than "I will."

Of course, the immediate crisis has passed, but there's still the same sense of unpredictability dogging our thinking. Best to concentrate on the current quarter, because who knows what job I'll have next year. Best to pass that test, because what I actually learn won't be worth much ten years from now anyway.

How can we reverse this?

It's a social problem, but I think it may also herald a mental one—which I imagine as a sort of mental diabetes. Most of us grew up reading books (at least occasionally) and playing with noninteractive toys that required us to make up our own stories, dialogue, and behavior for them. But today's children are living in an information-rich, time-compressed environment that often seems to stifle a child's imagination rather than stimulate it. Being fed so much processed information—video, audio, images, flashing screens, talking toys, simulated action games—is like being fed too much processed, sugar-rich food. It may seriously mess up children's informational metabolism—their ability to process information for themselves. Will they be able to discern cause and effect, put together a coherent story line, think scientifically, read a book with a single argument rather than a set of essays?

I don't know the answers, but these questions are worth thinking about, for the long term.

James J. O'Donnell

THE CLASSICIST AND CULTURAL HISTORIAN James J. O'Donnell is provost of Georgetown University and the author of *Avatars of the Word: From Papyrus to Cyberspace* and *Augustine: A New Biography*.

What do I believe is true even though I cannot prove it? The question has a double edge and thus needs two answers.

First, and most simply: Everything. In a strict Popperian reading, all the things I "know" are only propositions I have not yet falsified. They are best estimates—hypotheses that so far make sense of all the data I possess. I cannot prove that my parents were married on a certain day in a certain year, but I claim to "know" that date quite confidently. Sure, there are documents, but in fact in their case there are different documents, presenting two different dates. I recall the story my mother told to explain that, and I believe it, but I cannot prove I am right. I know Newton's laws—and, indeed, believe them—but I also now know their limitations and imprecisions and suspect that more surprises lurk in the future.

But that's a generic answer and not much in the forward-looking, optimistic spirit that animates the *Edge* Question. So let me propose this challenge to practitioners of my own historical craft. I believe there are in principle better descriptions and explanations for the development and sequence of human affairs than historians are capable of providing. We draw our data mainly from witnesses, who share our mortality—and, for that matter, often our limited viewpoint. And so we explain history in terms of human choices and the behavior of organized social units. The rise of Christianity, say, or the Norman Conquest seem to us to be events we can explain, and we explain them on the human scale. But it might be that events can be better explained on a much larger time scale or a much smaller scale of behavior. An outright materialist could argue that all my acts, from the day of my birth, have been determined by genetics and environment. It was fashionable a generation ago to argue a Freudian grounding for Luther's revolt, but in principle it could as easily be true and (if we could know it) more persuasive to demonstrate that his acts were determined at the molecular and submolecular level.

The problem here is that we are far from being able to outline such a theory, much less make it persuasive—or even comprehensible. Understanding just one other person's life in such microscopic detail would take longer than the investigator's lifetime.

So what is to be done? Of course, historians will constantly struggle to improve their techniques and tools. (The advance of dendrochronology—dating wood by the tree rings and thus dating buildings and other artifacts more accurately than ever before—is one example of the way in which technological progress can tell us things we never knew.) But we will also continue to read and write stories in the old style, because stories are the way human beings most naturally make sense of their world. An awareness of the powerful possibility of other orders of description and explanation should at least teach us some humility and give us thoughtful pause when we are tempted to insist too strongly on one version of history—the one we happen to think is true. Even a Popperian could see this kind of intuition as beneficial.

Jean Paul Schmetz

THE ECONOMIST Jean Paul Schmetz, formerly managing director of Burda Digital, a subsidiary of the Burda Media Group, now runs a hedge fund he started in 1998.

When considering the *Edge* Question, one has to remember the basis of the scientific method: formulating hypotheses that can be disproved. Those hypotheses that are not disproved can be believed to be true until disproved. Since it is more glamorous for a scientist to formulate hypotheses than it is to spend years disproving existing ones proposed by other scientists, and unlikely that someone will spend time and energy trying to disprove his or her own statements, our body of scientific knowledge is surely full of hypotheses that we believe to be true but will eventually be proved false.

So I turn the question around: What scientific ideas that have not been disproved do you believe are false?

As a theoretical economist, I believe that most ideas taught in Economics 101 will be proved false someday. Most of them would already have been officially defined as false, were economics as stringent as the hard sciences, but because of a lack of better hypotheses they are still widely accepted and used in economics and in general commentary. Eventually someone will come up with new ones explaining and predicting economic reality in a way that will render most existing economic beliefs false.

Nassim Nicholas Taleb

NASSIM NICHOLAS TALEB is an essayist, belletrist, and practitioner of uncertainty (i.e., mathematical trader) who focuses on the attributes of unexpected events, extreme deviations from the regular, and our consequent inability to forecast. He is the author of *Fooled by Randomness: The Hidden Rule of Chance in Life and in the Markets.*

We are good at fitting explanations to the past, all the while living in the illusion that we understand the dynamics of history.

I believe there is a severe overestimation of knowledge in what I call the "ex-post" historical disciplines, meaning almost all of social science (economics, sociology, political science) and the humanities—everything that depends on the nonexperimental analysis of past data. I am convinced that these disciplines do not provide much understanding of the world—or even of their own subject matter. Mostly, they fit a narrative that satisfies our desire (even need) for a story. The implications defy conventional wisdom: You do not gain much by reading the newspapers, history books, analyses, economic reports; all you get is misplaced confidence about what you know. The difference between a cab driver and a history professor is only one of degree; the latter is probably better at expressing himself.

In economics and finance, for example, there are plenty of experts (many of whom make more than a million dollars a year) who publish forecasts for the benefit of their clients. Just check their forecasts against the outcome. Their projections fare hardly better than random, meaning that their "stories" are convincing but do not seem to help you any more than listening to that cab driver would. Nor will a close reading of the newspapers make the slightest difference to your understanding of what the economy or the markets will do. Tests done by financial empiricists in the 1960s on the effect of the news on prices came to the same conclusion. If you look closely at the data, you will find that people tend to anticipate (though poorly) the regular fluctuations but miss out on the large deviations, which have a disproportionately large impact on the total outcome.

I am convinced, yet cannot prove quantitatively, that such overestimation of our knowledge can be generalized to any sort of narrative based on past information and lacking experimental verification. The economists got caught because we have data and means to check the quality of their knowledge; the historians, the news analysts, the biographers, the pundits can all hide a little longer. It is said that "The wise see things coming." To me, the wise are those who know they cannot see things coming.

Simon Baron-Cohen

SIMON BARON-COHEN is a professor of developmental psychopathology at the University of Cambridge and a fellow of Trinity College, Cambridge. He is also a director of the Autism Research Centre in Cambridge. He is the author of several books, including *The Essential Difference: Men, Women, and the Extreme Male Brain*.

I am not interested in ideas that cannot in principle be proved or disproved. I am as capable as the next guy of believing in an idea that is not yet proved, as long as it can in principle be proved or disproved. In my chosen field of autism, I believe that the cause will turn out to be assortative mating of two hyper-systemizers. I believe this because we already have three pieces of the jigsaw puzzle:

1. Fathers of children with autism are more likely to work in the engineering field, compared to fathers of children without autism. (Note that engineering is a chosen example because it involves strong systematizing. But other related scientific and technical fields, such as math or physics, would have been equally good examples to study.)

2. Grandfathers of autistic children—on both sides of the

family—are also more likely to work in the engineering field, compared to grandfathers of children without autism.

3. Both mothers and fathers of children with autism are super-fast at the embedded-figures test, a task requiring analysis of patterns and rules.

We have had these pieces of the puzzle since 1997, in the scientific literature. They do not yet prove the assortative-mating theory; they simply point to it as being highly likely. The causes of autism are probably complex, including—at the very least—multiple genes interacting with environmental factors, but the assortative-mating theory may describe some contributing factors.

Direct tests of the theory are needed. I will be the first to give up this idea if it is proved wrong, since I'm not in the business of holding onto wrong ideas. But I won't give up the idea simply because of its unpopularity with certain groups (such as those people who want to believe that the cause of autism is purely environmental). I will hold onto the idea until it has been properly tested. Popperian science is about being able to let go of an idea when the evidence goes against it, but it is also about being able to hold onto an idea until the evidence has been collected, if you have enough reasons to believe it might be true.

Kevin Kelly

KEVIN KELLY helped launch *Wired* magazine in 1993, and served as its executive editor until January 1999. He is now senior maverick for *Wired*. His latest book is *Cool Tools 2003.1*.

The orthodoxy in biology states that every cell in your body contains exactly the same DNA. It's your identity, your indelible fingerprint, and since all the cells in your body have been duplicated from your initial unique stem cell, these zillions of offspring cells all maintain your singular DNA sequence. It follows, then, that when you submit a tissue sample for genetic analysis, it doesn't matter where in your body it comes from. Normally technicians grab some from the easily accessible parts of your mouth, but they could just as well take some from your big toe, or your liver, or an eyelash, and get the same results.

I believe, but cannot prove, that the DNA in your body (and in the bodies of all living organisms) varies from part to part. I make this prediction based on something we know about biology, which is that nature abhors uniformity. Nowhere else in nature do we see identity maintained to such exactness. Nowhere else is there such fixity.

I do not expect intra-soma variation to diverge very much.

Indeed, the genetic variation among individual humans is relatively mild, among the least of all animals, so the diversity within a human body is unlikely to be greater than it is among human bodies—although that may be possible. More likely, intra-soma variation will be less than racial diversity but greater than zero.

A few biologists already know (even if most of the public doesn't) that the full sequence of DNA in your cells changes over time, since your chromosomes are shortened each time they divide in growth. Because of a bug in the system, DNA is unable to duplicate itself when it gets to the very tip of its chain, so at each cell division it winds up a few hundred bases short. This slight reduction after each of the cell's scores of divisions is currently seen as the chief culprit in cell death, and thus your own death. But the variation I believe is happening is more fundamental. My guess is that DNA mutates in a population of the cells in your body, much as it does in a population of bodies.

The consequences are more than just curious. At the trivial end, if my belief is true, it would matter where in your body a sample of your DNA is taken. And it would also matter *when* your DNA is sampled, as this variation could change over time. If my belief is true, this variation might have some effect on locating the correct seminal cells for growing replacement organs and tissues.

While I have no evidence for my belief right now, it is a provable assertion. It will be shown to be true or false as soon as we have ubiquitous, cheap, full-genome sequences at discount mall prices. That is, pretty soon. I believe that once we have a constant reading of our individual full DNA (many times over our lives), we will have no end of surprises. I would not be surprised to discover that pet owners accumulate some tiny fragments of their pets' DNA, which is somehow laterally transferred via viruses to their own cellular DNA. Or that dairy farmers amass noticeable fragments of bovine DNA. Or that the DNA in our limbs somehow drifts genetically in a "limby" way, distinct from the variation in the cells in our nervous system.

But all these considerations are minor compared with a possible major breakthrough in understanding. We have a pretty good idea of how the selection in natural selection works: Less fit organisms die. But when it comes to understanding how variation arises in Darwinian evolution, all we can say is "random mutation," which is another way of saying, "We don't know exactly." If there is intra-somatic variation, and if we could easily observe it via constant full-genome sequencing, then we might be able to figure out exactly how a mutation occurs and whether there are patterns to them, and to what extent such variation is induced or influenced by the body or the environment—all ideas that currently challenge the accepted Darwinian wisdom that the body does not directly influence the genetic makeup of a cell. Monitoring genetic drift within a body may be a window into the origins of mutation itself.

Even if these larger ideas don't pan out, the simple fact that DNA in each cell of your body is not 100-percent identical would be worth investigating. Such a fact would be a surprise, except to me.

Martin Nowak

MARTIN NOWAK is a professor of mathematics and biology at Harvard University and director of its Program for Evolutionary Dynamics. He is interested in all aspects of mathematical biology; in particular, he works on the dynamics of infectious diseases, cancer genetics, the evolution of cooperation and human language. He is coauthor (with Robert May) of *Virus Dynamics*.

I believe the following aspects of evolution to be true, without knowing how to turn them into (respectable) research topics.

Important steps in evolution are robust. Multicellularity evolved at least ten times. There are several independent origins of eusociality. There were a number of lineages leading from primates to humans. If our ancestors had not evolved language, somebody else would have.

Cooperation and language define humanity. Every special trait of humans is a derivative of language.

Mathematics is a language and therefore a product of evolution.

Tom Standage

TOM STANDAGE is the technology editor of *The Economist* and the author of several histories of science and technology, including *A History of the World in Six Glasses*.

I believe that the radiation emitted by mobile phones is harmless.

My argument is based not on the scientific evidence (because there isn't much and what little there is has either found no effect or is statistically dubious) but on a historical analogy to previous scares—about overhead power lines and cathode-ray computer monitors (VDUs). Both were also thought to be dangerous, yet years of research—decades, in the case of the power lines—failed to find conclusive evidence of harm.

Mobile phones seem to me to be the latest example of what has become a familiar pattern: Anecdotal evidence indicates that a technology may be harmful, and no matter how many studies fail to find evidence of harm, there are always calls for more research.

The underlying problem, of course, is the impossibility of proving a negative. During the fuss over GM (genetically modified) crops in Europe, there were repeated calls for proof that GM technology was safe. Similarly, in the aftermath of the BSE (bovine spongiform encephalopathy) scare in Britain, scientists were repeatedly asked for proof that British beef was safe to eat. But you cannot prove that something has no effect. Absence of evidence is not evidence of absence. All you can do is look for evidence of harm. If you don't find it, you can look again. If you still fail to find it, the question is still open. "Lack of evidence of harm" means both "Safe as far as we can tell" and "We still don't know if it's safe or not." Scientists are often unfairly accused of logic-chopping when they point this out.

I expect that mobile phones will turn out to be the latest in a long line of technologies that have raised health concerns subsequently deemed unwarranted. In the nineteenth century, telegraph wires were accused of affecting the weather and railway travel was believed to cause nervous disorders. The irony is that since my belief that mobile phones are safe is based on a historical analysis, I am on no firmer ground scientifically than those who believe mobile phones are harmful. Still, I believe they're safe, though I can't prove it.

Steven Giddings

STEVEN GIDDINGS is a theoretical physicist at the University of California, Santa Barbara.

I believe that black holes do not—as Stephen Hawking argued long ago—destroy information, thereby violating quantum mechanics. The reason is that strong gravitational effects undermine the statement that degrees of freedom inside and outside a black hole are independent of each other. This, then, would be the resolution of what has been called "the black hole information paradox."

On the first point, I am far from alone. Many string theorists and others now believe that black holes don't destroy information. Hawking himself recently announced that he believes this and has conceded a famous bet, but he has not yet published work containing a specific statement as to where his original logic went wrong.

The second point I believe but cannot yet prove to the point of convincing many of my colleagues. While it is widely believed that Hawking's early conclusion was wrong, there is disagreement over where exactly his calculation failed, and none of the arguments have clearly identified this point of failure—physicists have struggled with this paradox for more than three decades. If black holes do emit information instead of destroying it, this probably comes from a breakdown of what is known in physics as locality, the notion that phenomena at widely separated points cannot instantaneously influence each other. David Lowe, Joseph Polchinski, Leonard Susskind, Larus Thorlacius, and John Uglum have argued that a possible mechanism for this locality violation involves formation of long strings near the black hole horizon. Gary Horowitz and Juan Maldacena have argued that the singularity at the center of a black hole must be a unique state, in effect squeezing information in a ghostly way so that it emerges in the Hawking radiation. And others have made other suggestions.

But I believe (and my former student Matthew Lippert and I have published arguments) that the breakdown of locality that invalidates Hawking's work involves strong gravitational physics that makes it inconsistent to think of separate and independent degrees of freedom inside and outside a black hole. The assumption that these degrees of freedom are separate was fundamental to Hawking's original work. Our argument for where it went wrong has a satisfying generality that mirrors the generality of Hawking's argument: Neither depends on the specifics of what kind of matter exists in the theory.

We base our argument on a principle we call "the locality bound." This is a criterion for the independence of physical degrees of freedom. (In technical language, it entails the vanishing of commutators of corresponding operators.) Roughly, a degree of freedom corresponding to a particle at position x with momentum p and another at y with momentum q will be independent only if the separation between x and y is large enough so that the particles are both outside a black hole that would form from their mutual energy, but otherwise they fail to be independent. I believe that this is the beginning of a general criterion (which will ultimately be more precisely formulated) for where locality fails in physics. It could be the beginning of a deeper understanding of holography, which suggests that there is an equally fundamental and equivalent two-dimensional reality underlying our observed three-dimensional reality, and it should be relevant to black-hole physics because of the large relative energies of the Hawking radiation emitted by a black hole and the degrees of freedom falling into a black hole. But this is not fully proved. Yet.

Alexander Vilenkin

THE PHYSICIST Alexander Vilenkin is the director of Tufts University's Institute of Cosmology and coauthor (with E. P. S. Shellard) of *Cosmic Strings and Other Topological Defects*.

There are good reasons to believe that the universe is infinite.

If so, it contains an infinite number of regions of the same size as our observable region, which is 80 billion light-years across. It follows from quantum mechanics that the number of distinct histories that could occur in any of these finite regions in the finite time since the Big Bang is finite. (By "history," I mean not just the history of civilization but everything that happens, down to the atomic level.) The number of possible histories is fantastically large—it has been estimated as 10 to the power 10^{150}—but the important point is that it is finite.

Thus, we have an infinite number of regions like ours and only a finite number of histories that can play out in them. It follows that every possible history will occur in an infinite number of regions. In particular, there should be an infinite number of regions with histories identical to ours. So if you are not satisfied with the result of a presidential election, don't despair: Your candidate has won on an infinite number of Earths.

This picture of the universe robs our civilization of any claim to uniqueness: Countless identical civilizations are scattered in the infinite expanse of the cosmos. I find this rather depressing, but it is probably true.

Another thing I believe to be true but cannot prove is that our part of the universe will eventually stop expanding and will recollapse to a Big Crunch. But this will happen no sooner than 20 billion years from now, and probably much later than that.

Lawrence M. Krauss

LAWRENCE M. KRAUSS, the Ambrose Swasey Professor of Physics and director of the Center for Education and Research in Cosmology and Astrophysics at Case Western Reserve University, investigates the relations between quantum phenomena at fundamental scales and cosmology. He is the author of a number of science books for the general public, including *The Physics of Star Trek* and *Hiding in the Mirror: The Mysterious Allure of Extra Dimensions*.

I believe that our universe is not unique.

As science has evolved, our place within the universe has continued to diminish in significance: First it was felt that Earth was the center of the universe, then that the sun was the center, and so on. We now realize that we are near the edge of a galaxy that is itself located nowhere special in a large, potentially infinite universe full of other galaxies. Moreover, we know that even the stars and the visible galaxies are themselves but an insignificant bit of visible pollution in a universe otherwise dominated by stuff that doesn't shine. Dark matter outnumbers normal matter by a factor of ten, and now we have discovered that even matter (dark or not) is relatively insignificant. Empty space contains more than twice as much energy as that associated with all the matter, including the dark matter, in the universe.

Further, as we ponder the origin of our universe and the nature of the strange, dark energy that dominates it, every plausible theory I know of suggests that the Big Bang that created our visible universe was not unique. There are likely to be a large and possibly infinite number of other universes out there, some of which may be experiencing Big Bangs as I write, and some of which may have already collapsed inward into Big Crunches. From a philosophical perspective, this may be welcome news to those who find a universe with a definite beginning but no definite end dissatisfying: In the "metaverse," or "multiverse," things may seem much more uniform in time.

But philosophy aside, the existence of many different, causally disconnected universes—regions we will never be able to communicate directly with and thus forever out of reach of direct empirical verification—may have significant impact on our understanding of our own universe. Their existence may help explain why our own universe has certain otherwise unexpected features, because in a metaverse with a possibly infinite number of different universes, which may vary in their fundamental features, it could be that life like our own would evolve only in universes with a special set of characteristics.

Whether or not this anthropic argument is necessary to understand our universe—and I hope it isn't—I find it satisfying to speculate that not only are we not in a particularly special place in our universe but that our universe itself may be insignificant on a larger cosmic scale. The idea represents perhaps the ultimate Copernican Revolution.

John D. Barrow

JOHN D. BARROW is a professor of mathematical sciences in the Department of Applied Mathematics and Theoretical Physics, Cambridge University, and the author of several books on cosmology, including *The Infinite Book: A Short Guide to the Boundless, Timeless, and Endless*.

I believe but cannot prove that our universe is infinite in size, finite in age, and just one among many. Not only can I not prove that, but I believe that these statements will prove to be unprovable in principle and we will eventually hold that principle to be self-evident.

Paul J. Steinhardt

THE THEORETICAL PHYSICIST Paul J. Steinhardt is the Albert Einstein Professor in Science at Princeton University. His research spans problems in particle physics, astrophysics, cosmology, and condensed matter physics.

I believe that our universe is not accidental, but I cannot prove it.

Historically, most physicists have shared this point of view. For centuries, most of us have believed that the universe is governed by a simple set of physical laws—laws that are the same everywhere—and that these laws derive from a simple unified theory.

However, in the last few years, an increasing number of my most respected colleagues have become enamored of the anthropic principle—the idea that there is an enormous multiplicity of universes with widely different physical properties and that the properties of our particular, observable universe arise from pure accident. The features of our universe happen to be compatible with the evolution of intelligent life, but otherwise there is nothing remarkable about it. The change in attitude arises in part from the failure (so far) to find a unified theory that predicts our universe as the unique possibility. According to some recent calculations, superstring theory, the current best hope for a unified theory, allows an exponentially large number of different universes, most of which look nothing like our own. String theorists have turned to the anthropic principle for salvation.

Frankly, I view this as an act of desperation. I don't have much patience with the anthropic principle: The concept is, at heart, nonscientific. A proper scientific theory is based on testable assumptions and judged by its predictive power. The anthropic principle makes an enormous number of assumptions—regarding the existence of multiple universes, a random creation process, probability distributions that determine the likelihood of various features, and so on—none of which are testable, because they entail hypothetical regions of spacetime that are forever beyond the reach of observation. As for predictions, there are few if any. In the case of string theory, the anthropic principle is invoked only to explain known observations, not to predict new ones. (In some versions of the anthropic principle, in which predictions are made, the predictions have proved wrong: For example, recent evidence for a cosmological constant is said to have been anticipated by anthropic argument; however, the observed value does not agree with the anthropically predicted value.)

I find the desperation especially unwarranted since I see no evidence that our universe arose in a random process. Quite the contrary: Recent observations and experiments suggest that our universe is simple. The distribution of matter and energy is remarkably uniform. The hierarchy of complex structures, ranging from galaxy clusters to subnuclear particles, can be described in terms of a few dozen elementary constituents and less than a handful of forces, all related by simple symmetries. A simple universe demands a simple explanation. Why do we need to postulate an infinite number of universes with all sorts of different properties just to explain our own?

Of course, my colleagues and I expect to arrive at further reductionism. The current failure of string theory to find a unique universe may simply be a sign that our understanding of string theory is still immature (or perhaps that string theory is wrong). Decades from now, I hope physicists will be pursuing once again their dream of a truly scientific "final theory" and will look back at the current anthropic craze as millennial madness.

Lee Smolin

LEE SMOLIN is a founding member and research physicist at the Perimeter Institute for Theoretical Physics, in Waterloo, Ontario. A prominent contributor to the subject of quantum gravity, he is also the author of *The Life of the Cosmos* and *Three Roads to Quantum Gravity*.

I am convinced that quantum mechanics is not a final theory. I believe this because I have never encountered an interpretation of the present formulation of quantum mechanics that makes sense to me. I have studied most of them in depth and thought hard about them, and in the end I still can't make real sense of quantum theory as it stands. Among other issues, the measurement problem seems impossible to resolve without changing the physical theory.

Quantum mechanics must then be an approximate description of a more fundamental physical theory. There must then be *hidden variables*, which are averaged to derive the approximate, probabilistic description that is quantum theory. We know from the experimental falsifications of the Bell inequalities that any theory that agrees with quantum mechanics on a range of experiments where it has been checked must be nonlocal. Quantum mechanics is nonlocal, as are all proposals for replacing it with something that makes more sense. So any additional hidden variables must be nonlocal. But I believe we can say more. I believe that the hidden variables represent relationships between the particles we do see—relationships that are hidden because they are nonlocal and connect widely separated particles.

This fits in with another core belief of mine, deriving from general relativity, which is that the fundamental properties of physical entities are a set of relationships that evolve dynamically. There are no intrinsic, nonrelational properties, and there is no fixed background—such as Newtonian space and time—that exists just to give things properties.

One consequence of this is that the geometry of space and time is also only an approximate, emergent description, applicable only on scales too large for us to see the fundamental degrees of freedom. The fundamental relations are nonlocal with respect to the approximate notion of locality that emerges at the scale where it becomes sensible to talk about things located in a geometry.

Putting these notions together, we see that quantum uncertainty must be a residue of the resulting nonlocality, which restricts our ability to predict the future of any small region of the universe. H-bar, the fundamental constant of quantum mechanics that measures the quantum uncertainty, is related to N, the number of degrees of freedom in the universe. A reasonable conjecture is that h-bar is proportional to the inverse of the square root of N.

But how are we to describe physics, if it is not in terms of things moving in a fixed spacetime? Einstein struggled with this, and my only answer is the one he came to near the end of his life: Fundamental physics must be discrete, and its description must be in terms of algebra and combinatorics.

And what of time? I have also been unable to make sense of

any of the proposals to do away with time as a fundamental aspect of our description of nature. So I believe in time, in the sense of causality. I also doubt that the Big Bang is the beginning of time. I strongly suspect that our history extends backward before the Big Bang.

Finally, I believe that in the near future we will be able to make predictions based on these ideas which will be tested in real experiments.

Anton Zeilinger

ANTON ZEILINGER, a professor of physics at the University of Vienna, has conducted quantum teleportation experiments and quantum interference experiments with "buckyball" molecules, the largest objects ever to have demonstrated quantum phenomena. His next goal is to extend the validity of quantum phenomena experimentally to the realm of even larger objects, perhaps even to life itself.

What I believe but cannot prove is that quantum physics requires us to abandon the distinction between information and reality.

Why do I believe this? Because it is impossible to make an *operational* distinction between reality and information. Whenever we make any statement about the world, about an object, about a feature of an object, we are making a statement about the information we have. And whenever we make scientific predictions, we make statements about information we hope to attain in the future. One might therefore be tempted to believe that everything is just information; the danger here is solipsism and subjectivism. But we know (even though we cannot prove it) that reality is "out there." For me, the strongest argument for a reality independent of myself is the randomness of the individual quantum event—the decay of a radioactive atom, for instance. There is no hidden reason that a given atom should decay at the very instant it does so.

So, if reality exists, and if we will never be able to make an operational distinction between reality and information, it would seem that reality and information are one and the same. We need a concept that encompasses both.

This is the message of the quantum. It is the natural extension of the so-called Copenhagen interpretation, which holds that we must never assign features to an object without having actually observed them. Once you adopt the notion that reality and information are the same, all quantum paradoxes and puzzles—like the measurement problem, or the nine lives of Schrödinger's cat—disappear. Yet the price of reconciliation is high. If my hypothesis is true, many questions become meaningless. There is no sense asking what is "really"going on out there. Schrödinger's cat is neither dead nor alive unless we obtain information about its state.

By the way, I also believe that the day will come when we learn to overcome "decoherence" and to observe quantum phenomena outside the shielded environment of laboratories. I hope that (unlike the unexamined cat) I will be alive when this happens.

Gregory Benford

GREGORY BENFORD is a professor of plasma physics and astrophysics at the University of California at Irvine and a prolific writer of science fiction novels, the most recent of which is *The Sunborn*, a sequel to the 1999 *The Martian Race*. He is also the author of *Deep Time: How Humanity Communicates Across Millennia*.

Why is there scientific law at all? I have a possible answer, but as yet no proof of it.

We physicists explain the origin and structure of matter and energy but not the origin of the laws behind them. Does the idea of causation apply to where the laws themselves came from? We have narrowed the range of field theories that can yield the Big Bang universe we live in, but why do the laws that govern it seem constant in time and always at work?

One can imagine a universe in which laws are not truly lawful. Talk of miracles does just this, invoking God to make things work. Physics aims to find the laws instead, and hopes that they will be uniquely constrained, as when Einstein wondered whether God had any choice when He made the universe. One fashionable escape hatch from the problem of choice is to assert that there are infinitely many universes, each sealed off from the others, which can obey any sort of law you can imagine, with all sorts of parameters and assumptions. This "multiverse" view represents the failure of our grand agenda and seems to me contrary to the prescribed simplicity of Occam's Razor, solving our lack of understanding by multiplying unseen entities into infinity.

Perhaps it is a similar philosophical failure of imagination to think, as I do, that when we see order there is usually an ordering principle. But what can constrain the nature of physical law? Natural selection gave us our ornately structured biosphere, and perhaps a similar evolutionary principle operates in the genesis of universes. Our universe may have arisen from selection for intelligences that can make fresh universes—perhaps in high-energy physics experiments, or near black holes (as Lee Smolin has supposed), where spacetime gets contorted into plastic forms that can make new spacetimes. An Ur-universe that had intelligence could make others, and this reproduction, with perhaps slight "genetic" variation, would drive the evolution of physical law. The astrophysicist Edward Harrison has had similar ideas.

Selection arises because only firm laws can yield constant, benign conditions to form new life. Once life-forms realized this, they could intentionally make more smart universes with the right fixed laws to produce ever grander structures. There might be observable consequences of this prior evolution. If it did occur, then we are an inevitable consequence of the universe, mirroring intelligences who came before, in some earlier universe, and deliberately chose to create further sustainable order. The fitness of our cosmic environment is then no accident. If we continue to find signs of cosmological fine-tuning, is it evidence for such views?

Rudy Rucker

RUDY RUCKER is a mathematician, computer scientist, cyberpunk pioneer, and novelist. He is the author of *Infinity and the Mind* and *The Lifebox, the Seashell, and the Soul.*

I'd like to propose a modified Many Universes theory. Rather than saying that every possible universe exists, I'd say that there is a sequence of possible universes, akin to the drafts of a novel. We're living in a draft version of the universe, and there is no final version. The revisions never stop.

From time to time, it's possible to be aware of this. In particular, when you relax and stop naming things and forming opinions, your consciousness spreads out across several drafts of the universe. Things don't need to be particularly one way or the other until you pin them down.

Each draft, each spacetime, each sheet of reality is itself rigorously deterministic. There is no underlying randomness in the world; instead, we have a great web of synchronistic entanglements, with causes and effects flowing forward and backward through time. The start of a novel matches its ending; the past matches the future. Changing one thing changes everything. If we know everything about the Now moment, we know the entire past and future.

Explaining any given draft of the universe thus becomes a

matter of explaining the contents of a single Now moment of that draft. This in turn means that we can view the evolution of the successive drafts as an evolution of different versions of a particular Now moment. As Scarlett O'Hara's climactic scene with Rhett Butler is repeatedly rewritten, all the rest of *Gone With the Wind* changes to match.

And this evolution, too, can be deterministic. We can think of there being two distinct deterministic rules: a physics rule and a metaphysics rule. The physics rule consists of time-reversible laws that grow the Now moment downward and upward to fill the entire past and future of spacetime. And we invoke the metaphysics rule to account for the contents of the Now moment. The metaphysics rule is deterministic but not reversible; it grows sideways across a dimension we might call "paratime," turning some simple seed into the space-filling pattern found in the Now.

The metaphysics rule is . . . what? One possibility is that it's something quite simple, perhaps as simple as an eight-bit cellular-automaton rule generating complex-looking patterns out of pure computation. Or perhaps the metaphysics rule is like the mind of an author creating a novel, searching out the best word to put down on paper next, peering into alternative realities. Or yet again, the big metaphysics rule in the sky could be the One Cosmic Mind, the Big Aha!, the Eternal Secret, living in the spaces between your thoughts.

Carlo Rovelli

CARLO ROVELLI is a physicist at the Centre de Physique Théorique, in Marseille. He is a senior member of the Institut Universitaire de France, a professor at the Université de la Méditerranée, and an affiliated professor of history and the philosphy of science at the University of Pittsburgh. He is also the author of *Quantum Gravity*.

I am convinced, but cannot prove, that time does not exist; that is, there is a consistent way of thinking about nature that makes no use of the notions of space and time at the fundamental level. And I believe that this way of thinking will turn out to be the useful and convincing one. I think the notions of space and time will turn out to be useful only within some approximation. They are similar to notions like "the surface of the water," which loses meaning when we describe the dynamics of the individual atoms forming water and air; on the smallest of scales, there isn't really any surface there. I am convinced that space and time are, like the surface of the water, convenient macroscopic approximations—flimsy and illusory, screens that our minds use to organize reality.

In particular, I am convinced that time is an artifact of the approximation by which we disregard most of the degrees of freedom of reality. Thus "time" is, in a sense, the reflection of our ignorance.

I am also convinced but cannot prove that there are no objects, only relations. By this I mean that there is a consistent way of thinking about nature that refers only to interactions between systems and not to states of or changes in individual systems. Likewise, I believe that this way of thinking about nature will turn out to be the useful and natural one in physics.

Beliefs that one cannot prove are often wrong (as demonstrated by the contradictory beliefs in these pages). But they are also often healthy, and they are essential in science. Here is a good example from twenty-four centuries ago: Socrates, in Plato's *Phaedo*, says, "I know not that the art of Glaucus could prove the truth of my tale, which I myself should never be able to prove . . . [but] my conviction is that the earth is a round body. . . ."

Finally, I also believe but cannot prove that we humans have the collaborative instinct. This instinct will eventually prevail over the shortsighted, egoistic, and aggressive instinct that produces exploitation and war. Collaboration has already given us long periods of peace and prosperity. Ultimately it will lead to a planet without countries, without wars, without patriotism, without religions, without poverty—and we will be able to share the world. Actually, I'm not sure I believe that I believe this, but I do want to believe that I do.

Jeffrey Epstein

JEFFREY EPSTEIN is a money manager and science philanthropist.

I believe that the mechanism for the human perception of time will be discovered. Almost another sense—the ability to distinguish past from present, in intervals long enough to convey a thought and create memories—will establish a new boundary for consciousness.

There will be found (in addition to entropy) a cost, or friction, for just moving through time. Steady states will be the classical limit. We will uncover the formula of time's relationship to life, which will be as unique as time's relationship to space.

Howard Rheingold

HOWARD RHEINGOLD is an electronic and computer-mediated communications expert and the author of several books, including *The Virtual Community* and *Smart Mobs: The Next Social Revolution*.

I believe that we humans, who know so much about cosmology and immunology, lack a framework for thinking about why and how humans cooperate. I believe that part of the reason for this is an old story we tell ourselves about the world: Businesses and nations succeed by competing well; biology is a war in which only the fit survive; politics is about winning; markets grow solely from self-interest. Still rooted in the zeitgeist of Adam Smith's and Charles Darwin's eras, the scientific, social, economic, and political stories of the past century have overwhelmingly emphasized the role of competition as a driver of evolution, progress, commerce, and society.

I believe that the outlines of a new narrative are becoming visible—a story in which cooperative arrangements, interdependencies, and collective action play a more prominent part, while that of (the essential but not all-powerful) competition and survival of the fittest shrinks just a bit.

Although new findings about the evolution of altruistic behavior and symbiotic relationships, new understandings of economic behavior derived from experiments in game theory and neuroeconomic research, sociological investigations of institutions for collective action, and computation-enabled technologies (such as grid computing, mesh networks, and online markets) have all provided important clues, I don't expect anyone to formulate an algorithm or a recipe for human cooperation. The complex interdependencies of human thought, behavior, and culture are doubtless equivalent to the limits that Werner Heisenberg found in physics and Kurt Gödel established for mathematics: Human social behavior is a complex adaptive system and thus not deterministic.

I believe that more knowledge than what we have now, together with a conceptual framework that is neither reductive nor theological, could lead to better designed economic and political policies and institutions. The institutional and conceptual barriers to mounting such an effort are as formidable as the methodological barriers. I am reminded of the problem faced in the 1950s by Doug Engelbart, the man who invented so much of today's mind-augmenting technologies (the graphic user interface, the mouse, hypertext, text editing, online group communications). He couldn't convince computer engineers, librarians, or public policy analysts that computing machinery could be used to enhance human thinking, as well as to perform scientific calculation and business-data processing. Nobody and no institution had ever thought about computing machinery that way, and older ways of thinking about what machines could be designed to do were inadequate. Engelbart had to construct "a conceptual framework for augmenting human intellect" before the various hardware, software, and interface designers could create the first personal computers and networks.

Useful new understandings of how humans cooperate (or

fail to cooperate) are of necessity an interdisciplinary task. The obvious importance of such an effort is no guarantee that it will be successfully accomplished. All our institutions for gathering and validating knowledge—universities, corporate research laboratories, foundations—reward and support specialization.

Jaron Lanier

JARON LANIER is a computer scientist, composer, and visual artist, probably best known for his work in virtual reality, a term he coined. His current research interests include real-time remote terascale processing, autostereo methods, and haptics.

My belief is that the potential for expanded communication between people far exceeds the potential of language and of all the other communication forms we already use.

Suppose for a moment that children in the future will grow up with an easy and intimate virtual-reality technology and that their use of it will become focused on invention and design instead of the consumption of pre-created holo-video games, surround movies, or other content.

Maybe these future children will play virtual musical-instrument-like things that cause simulated trees and spiders and seasons and odors and ecologies to spring up, just as manipulating a pencil causes words to appear on a page. If people grew up with a virtuosic ability to improvise the contents of a shared virtual world, a new sort of communication might also appear.

It's barely possible to imagine what a "reality conversation" would be like. Each person would be changing the shared world at the speed of language all at once, an image that suggests chaos, but often there would be a coherence that would indicate meaning. A kid becomes a monster, eats his little brother, who becomes a vitriolic turd, and so on.

This is what I've called postsymbolic communication, though it won't exist in isolation of, or in opposition to, symbolic communication techniques. It will be something different, however, and will expand what people can mean to each other.

Postsymbolic communication will be like a shared, waking-state, intentional dream. Instead of uttering the word "house," you will create a particular house and be able to walk into it; and instead of comprehending the category "house," you will peer into an apparently small bucket big enough to hold all the universe's houses, so that you can assess directly what they have in common. It will be a fluid form of experiential concreteness, providing expressive power similar to but divergent from that of abstraction.

Why care? The acquisition of postsymbolic communication will be a centuries-long adventure, an expansion of meaning, something beautiful, and a way to seek cool, advanced technology that focuses on connection instead of mere power. It will be a form of beauty that also enhances survivability. Since the drive for "cool tech" is unstoppable, the invention of provocative cool tech lovely enough to seduce the attention of young smart people away from arms races is a prerequisite to the survival of the species.

Some of the aforementioned examples (houses, spiders) apply to improvisations of the external environment, but postsymbolic communication might typically look a lot more like people morphing themselves into varied forms. Experiments have already been conducted with kids wearing special body suits and goggles and "turning into" triangles to learn trigonometry or into molecules to learn chemistry.

It's not just the narcissism of the young (and not so young) human mind or the primalness of the control of one's body that makes self-transformation compelling. Evolution, as generous as she ended up being with us humans, was stingy with potential means of expression. Compare us with the mimic octopus, which can morph into all sorts of creatures and objects and can animate its skin. An advanced civilization of cephalopods might develop words as we know them, but probably only as an adjunct to a natural form of postsymbolic communication.

We humans can control precious little of the world with enough agility to keep up with our thoughts and feelings: The fingers and the tongue are just about it. Symbols in language are a trick (or what programmers call a hack) that expands the power of little appendage wiggles to refer to all that we can't instantly become or create.

While we're confessing unprovable beliefs, here's another one: The study of the genetic components of pecking-order behavior, group-belief cues, and clan identification leading to inter-clan hostility will be the core of psychology and sociology for the next few generations, and it will become clear that we can't turn off or control these elements of human character without losing other qualities we love, like creativity. If this dark guess is correct, then the means to survival is to create societies with a huge variety of paths to success and a multitude of overlapping, intertwined clans and pecking orders, so that everyone can be a winner from equally valid individual perspectives. When the American experiment has worked best, it has approximated this level of variety. The virtual worlds of postsymbolic communication can provide the highest level of variety to satisfy the dangerous psychic inheritance I'm guessing we suffer as a species.

Implicit in the futures I am imagining is a solution to the software crisis. If children are breathing out fully realized creatures just as they form sentences today, there must be software present that isn't crashing and is marvelously flexible and responsive yet free of limiting preconceptions (which would simply be a revival of symbolism). Can such software exist? Ah! Another belief! My guess is that it can exist but not anytime soon. The only two good examples of software we have now are evolution and the brain, and they are both quite good, so why not be encouraged?

Marti Hearst

MARTI HEARST is an associate professor in the School of Information Management and Systems at UC Berkeley, with an affiliate appointment in the Computer Science Division. Her primary research interests are user interfaces and visualization for information retrieval, empirical computational linguistics, and text data mining.

I believe that the search problem is solvable. Advances in computational linguistics and user interface design will eventually enable people to find answers to any question they may have, as long as the answer is encoded in textual form and stored in a publicly accessible location. Advances in reasoning systems will be able to draw inferences in order to find answers not explicitly present in the existing documents.

Several recent developments prompt me to make this claim. First, computational linguistics (also known as natural language processing, or language engineering) has made great leaps in the last decade, primarily because of advances stemming from the availability of huge text collections, from which statistics can be derived. Today's language translation systems, for example, come almost entirely from statistical patterns extracted from text collections; they work as well as hand-engineered systems and will continue to improve.

Search engine companies also have enormous repositories of information about how people ask for information. This behavioral information can be used to build better search tools. For example, some spelling correction algorithms make use of how people correct erroneous spellings by observing pairs of serial queries: The second query is assumed to be a spelling correction if it is enough like the first. Patterns are then derived that convert various kinds of misspellings to their corrections.

Another development in the field of computational linguistics is the manual creation of enormous lexical ontologies, which are then used to build axioms and rules about language use. These modern ontologies, unlike their predecessors, are of a large enough and simple enough design to be useful, although this work is in the early stages. There are also many attempts to build such ontologies automatically, from large text collections; the most promising approach seems to be to combine the automated and the manual approaches. (I am skeptical about the hype surrounding the Semantic Web—it is very difficult to characterize concepts in a systematic way and even more so to force all the world's creators of information to conform to one schema.)

Finally, advances in user interface design are key to producing better search results. The search field has learned an enormous amount in the ten years since the Web became a major presence in society but—as is often noted in the field—the interface itself hasn't changed much: After all this time, we're still typing words into a blank box and then selecting from a list of results. I believe headway will be made in this area, most likely occurring in tandem with advances in natural language analysis.

Kai Krause

KAI KRAUSE has a doctorate in philosophy, a master's in image processing, a patent for interface concepts, a Clio for the first *Star Trek* movie, and a Davis Medal from the Royal British Photographical Society. He is the founder of a research lab called Byteburg in a 1,000-year-old castle above the Rhine River.

I have always felt, but cannot prove, that *Zen* is wrong. *Then* is right. Everything is not about the *now*, as in "the here and now," "living for the moment," and so on. On the contrary: I believe that everything is about the *before then* and the *back then*. It is about the anticipation of the moment and the memory of the moment, but not the moment.

In German, there is a beautiful little word for it: *Vorfreude*, which is a shade different from "delight" or "pleasure" or even "anticipation." It is the "pre-delight," the "before joy," or, as a little linguistic concoction, the "fore-fun." A single word captures the relationship of time, the pleasure of waiting for the moment to arrive, the *can't wait* moments of elation, of hoping for something, someone, some event to happen—whether it's on a small scale (like anticipating the arrival of a loved one or that moment in a piece of music or that sequence in a movie) or the larger versions (the expectation of a beautiful vacation, of the birth of a baby, of your acceptance of an Oscar).

We have been told by wise men, lamas, and maharishis that it is supposedly all about moments—to cherish the moment and never mind the continuance of time. But ever since childhood I have realized somehow that the beauty lies in the time before, the hope for, the waiting for, the imaginary picture painted in perfection of that instant in time. And then once it passes, in the blink of an eye, it will be the memory that stays with you, the reflection, the remembrance of that time.

Nothing ever is as beautiful as its abstraction seen through the rose-colored glasses of anticipation. The toddler's hoped-for Santa Claus on Christmas Eve turns out to be a fat guy with a fashion issue. Waiting for the first kiss can give you waves of emotional shivers up your spine, but when it actually happens it's a bunch of molecules colliding—a bit of a mess, really. In anticipation, the moment will be glorified by innocence, not knowing yet. In remembrance, the moment will be sanctified by memory filters, not knowing any more.

In the Zen version, trying to uphold the beauty of the moment in that moment is, in my eyes, a sad undertaking. Not so much because it can't be done: All manner of techniques have taught us how to be a happy human by mastering the art of it. But it implies, by definition, that all those other moments live just as much under the spotlight: the mundane, the lame, the gross, the everyday routines of dealing with life's mere mechanics.

In the Then version, it is quite the opposite: The long phases before and after last hundreds or thousands of times longer than the moment and drown out the everyday humdrum entirely.

Bluntly put: Spend your life in the eternal bliss of always having something to hope for, something to wait for, plans not realized, dreams not yet come true. Make sure you have new points on the horizon, that you deliberately create. And at the same time relive your memories, uphold and cherish them, keep them alive and share them, talk about them.

Make plans and take pictures.

I have no way of proving the rightness of such a lofty philosophical theory, but I greatly anticipate the moment that I might, and once I have done it I will most certainly never forget it.

Oliver Morton

OLIVER MORTON is a freelance writer, a contributing editor at *Wired*, and the author of *Mapping Mars*.

I've always found belief a bit difficult; most of what I believe to be true lies far beyond my ability to prove it. As far as knowledge goes, I'm a consumer and sometimes a distributor, not a producer. My beliefs are based on faith in other people and in processes and institutions.

The same is true for most of us. Those who can prove their beliefs in their own field of expertise still rely on faith in others when it comes to other fields. To continually acknowledge this would make every utterance tentative, encrust every concept with *ceteris paribus* clauses. But when faced with a question about our beliefs, the role of faith in people and social institutions has to be credited.

I suppose the real question is, What do I believe that I don't think *anyone* can prove? In answer, I'd put forward the belief that there is a future much better, in terms of reduced human suffering and increased human potential, than the present, and that one part of what will make it better is a greater, subtler knowledge of the world at large.

If I can't prove this, why do I believe it? Because it's better than believing the alternative. Because it provides a context for social and political action that would otherwise be futile. In this, it is an exhortatory belief. It is also, in part, a self-serving one, in that it suggests that when I try to clarify and disseminate knowledge (a description that makes me sound like the chef at a soup kitchen) I'm doing something in aid of that better future, if only a bit.

Besides the question of why, though, there's the question of how. And the answer to that one is "with difficulty." It is not an easy thing for me to make myself believe. But it is what I want to believe, and on my best days I do.

W. Daniel Hillis

THE PHYSICIST, computer scientist, and inventor W. Daniel Hillis pioneered the concept of parallel computers. He is currently the chairman of Applied Minds, Inc., a research and development company creating a range of products and services in software, entertainment, electronics, biotechnology, and mechanical design. He is also the author of *The Pattern on the Stone: The Simple Ideas That Make Computers Work*.

I know it sounds corny, but I believe that people are getting better. In other words, I believe in moral progress. It is not a steady progress, but there is a long-term trend in the right direction—a two-steps-forward-one-step-back kind of progress.

I believe, but cannot prove, that our species is passing through a transitional stage from being animals to being true humans. I do not pretend to understand what true humans will be like, and I expect that I would not understand it even if I met them. Yet I believe that our own universal moral sense is pointing us in the right direction and that it is the direction of our future.

I believe that 10,000 years from now, people (or whatever we are by then) will be more empathetic and more altruistic than we are. They will trust one another more, and for good reason. They will take better care of one another. They will be more thoughtful about the broader consequences of their actions. They will take better care of their future than we do of ours.

Martin E. P. Seligman

MARTIN E. P. SELIGMAN is Fox Leadership Professor of Psychology in the Department of Psychology at the University of Pennsylvania. His research focuses on positive psychology, learned helplessness, depression, and optimism and pessimism. The latest of his twenty books is *Authentic Happiness*.

The "rotten-to-the-core" assumption about human nature espoused so widely in the social sciences and the humanities is wrong. This premise has its origins in the religious dogma of original sin and was dragged into the secular twentieth century by Freud and reinforced by two world wars, the Great Depression, the cold war, and genocides too numerous to list. The premise holds that virtue, nobility, meaning, and positive human motivation generally are reducible to, parasitic upon, or compensations for what is really authentic about human nature: selfishness, greed, indifference, corruption, and savagery. The only reason I am sitting in front of this computer typing away rather than running out to rape and kill is that I am "compensated"—that is, zipped up, successfully defending myself against those fundamental underlying impulses.

In spite of its widespread acceptance in the religious and academic world, there is not a shred of evidence, not an iota of data, compelling us to believe the idea that nobility and virtue are somehow derived from negative motivation. On the contrary, I believe that evolution has favored both positive and negative traits; many niches have selected for morality, cooperation, altruism, and goodness, just as many have selected for murder, theft, self-seeking, and terrorism.

More plausible than the rotten-to-the-core theory of human nature is a dual-aspect theory: that the strengths and the virtues are just as basic to human nature as the negative traits are, and that negative motivation and emotion have been selected for in evolution. Evolution, after all, works through two processes: zero-sum-game survival struggles lubricated by negative emotion—anxiety, anger, and sadness—on the one hand, and sexual selection on the other, a positive-sum-game process that has favored virtue and is lubricated by positive emotion. These two overarching systems sit side by side in our central nervous system, ready to be activated (on the one hand) by privation and thwarting, or (on the other) by abundance and the prospect of growth and success.

Neil Gershenfeld

THE PHYSICIST Neil Gershenfeld is the director of MIT's Center for Bits and Atoms, which investigates the boundary between physical science and computer science, from molecular quantum computation to digital fabrication. He is the author of *FAB: The Coming Revolution on Your Desktop—from Personal Computers to Personal Fabrication.*

What do I believe is true even though I cannot prove it?

Progress.

The enterprise that employs me, seeking to understand and apply insight into how the world works, is ultimately based on the belief that this is a good thing to do. But it's something of a leap of faith to believe that that will leave the world a better place—the evidence to date is mixed for technical advances monotonically mapping onto human advances.

Naturally, this question has a technical spin for me. My current passion is the creation of tools for personal fabrication based on additive digital assembly, so that the uses of advanced technologies can be invented by their users. It's still no more than an assumption that that will lead to the making of more good things than bad things—but, like the accumulated experience indicating that democracy works better than monarchy, I have more faith in a future based on widespread access to the means for invention than in one based on technocracy.

Mihaly Csikszentmihalyi

MIHALY CSIKSZENTMIHALYI is a professor at the Drucker School of Management at Claremont Graduate University and the director of its Quality of Life Research Center. His books include the bestselling *Flow*, *The Evolving Self*, *Creativity*, and *Good Business*.

I can prove almost nothing I believe in. I believe the earth is round but I cannot prove it, nor can I prove that the earth revolves around the sun or that the naked fig tree in the garden will have leaves in a few months. I can't prove that quarks exist or that there was a Big Bang—all of these and millions of other beliefs are based on faith in a community of knowledge whose proofs I am willing to accept, hoping they will accept on faith the few measly claims to proof I might advance.

But now I realize, having read some of the other postings, that everyone else has assumed implicitly that the "you" in "even if you cannot prove it" refers not to the individual respondent but to the community of knowledge. It actually stands for "one" rather than for "you." That everyone seems to have understood this seems to me a remarkable achievement, a merging of the self with the collective that only great religions and profound ideologies occasionally achieve.

So, what do I believe that no one else can prove? Not much,

although I do believe in evolution, including cultural evolution, which means that I tend to trust ancient beliefs about good and bad, the sacred and the profane, the meaningful and the worthless—not because they are amenable to proof but because they have been selected over time and in different situations and therefore might be worthy of belief.

As to the future, I will follow the cautious weather forecaster who announces, "Tomorrow will be a beautiful day, unless it rains." I can see all sorts of potentially wonderful developments in human consciousness, global solidarity, knowledge, and ethics. However, there are about as many trends operating toward opposite outcomes: a coarsening of taste; a reduction to least common denominators; a polarization of property, power, and faith. I hope we will have the time and opportunity to understand which policies lead to which outcomes, and then that we will have the motivation and the courage to implement the more desirable alternatives.